Orbitals in Chemistry

Modern chemistry owes a great deal to two fundamental concepts, energy and probability, usually coupled to each other. One of the main examples of such a fruitful convergence is the field of quantum chemistry and orbital theory. The orbital concept provides the basic preparation in atomic and molecular structure theory needed for the understanding and interpretation of organic chemistry, inorganic chemistry and spectroscopy.

This text presents a unified and up-to-date discussion of the role of atomic and molecular orbitals in chemistry, from the quantum mechanical foundations to the recent developments (brief discussion of special systems such as solids and C_{60}, density-functional theory and Kohn–Sham orbitals) and applications (physical properties, reactivity, spectroscopy). The discussion is mainly qualitative, largely based on symmetry arguments. It is felt that a sound mastering of the concepts and qualitative interpretations is needed especially when students are becoming more and more familiar with numerical calculations based on atomic and molecular orbitals. The text is mathematically less demanding than most traditional quantum chemistry books but still retains clarity and rigour. The physical insight is maximized and abundant illustrations are used. The relationships between the more formal quantum-mechanical formalisms and the traditional descriptions of chemical bonding are critically established.

This book is of primary interest to undergraduate chemistry students and others taking courses of which chemistry is a significant part.

VICTOR M. S. GIL was born in Portugal on 13 March 1939. His Ph.D. was in NMR spectroscopy at the University of Sheffield in 1965. He is past Chairman of the Installation Committee and was first Rector of the University of Aveiro (Portugal) from 1973 to 1977. He is currently Professor of Chemistry at the University of Coimbra, Portugal, and Director of Explorations, the interactive science centre of Coimbra. His active area of research is NMR spectroscopy, in particular spectral parameters and molecular structure, conformational analysis, sugar chemistry and coordination compounds (vanadium, molybdenum, tungsten and uranium). He is author of over 100 scientific papers, co-author of a book on NMR spectroscopy and has written several papers on the teaching and learning of chemistry. He has also co-authored the current chemistry syllabus for Portuguese schools (from 8th grade to 12th grade) and is co-author of several books and software products for chemistry teaching in schools. He has taught more than 10 000 undergraduate students and has supervised 10 Ph.D. and M.Sc. theses along his career.

Orbitals in Chemistry
A Modern Guide for Students

VICTOR M. S. GIL
University of Coimbra, Portugal

CAMBRIDGE
UNIVERSITY PRESS

PUBLISHED BY THE PRESS SYNDICATE OF THE UNIVERSITY OF CAMBRIDGE
The Pitt Building, Trumpington Street, Cambridge, United Kingdom

CAMBRIDGE UNIVERSITY PRESS
The Edinburgh Building, Cambridge CB2 2RU, UK   http://www.cup.cam.ac.uk
40 West 20th Street, New York, NY 10011-4211, USA   http://www.cup.org
10 Stamford Road, Oakleigh, Melbourne 3166, Australia
Ruiz de Alarcón 13, 28014 Madrid, Spain

First published 2000

Printed in the United Kingdom at the University Press, Cambridge

Typeface Times 11/14pt.   System 3b2 6.03d [ADVENT]

A catalogue record for this book is available from the British Library

Library of Congress Cataloguing in Publication data

Gil, Victor M. S., 1939–
Orbitals in chemistry: modern guide for students/Victor M. S. Gil.
p. cm.
Includes bibliographical references.
ISBN 0-521-66167-6 (hb)
1. Molecular orbitals. I. Title.

QD461 G52 2000
541.2′8–dc21   99-461968

ISBN 0 521 66167 6 hardback
ISBN 0 521 66649 X paperback

Contents

541.28
61

Preface

Modern chemistry owes a great deal to two fundamental concepts, energy and probability, which appear frequently coupled to each other. Main examples of such a fruitful convergence are the second law of thermodynamics and entropy (in the field of chemical transformations) and quantum chemistry and orbitals (in the field of structure of atoms and their groupings and properties). Orbitals are the central subject of this book, presented in a unified and updated review. The discussion is mainly qualitative, largely based on symmetry arguments. It is felt that a sound mastering of the concepts and qualitative interpretations is needed especially when students are becoming more and more familiar with numerical calculations based on atomic and molecular orbitals. The text is mathematically less demanding than most traditional quantum chemistry books, but still retains clarity and rigour.

This is thus a chemically oriented, not too advanced book, written in an easy-going style. Without loss of correctness, the physical insight is maximized and abundant graphical illustrations are used. A large number of figures (127) and many diagrams help to make the book less dense and more instructive. The relationships between the more formally based quantum-mechanical formalisms and the traditional chemical descriptions of atomic and molecular structure are established and discussed in a critical manner.

A selected collection of problems (120) is included. Instead of listing the problems at the end of each chapter, they are presented at the right moment in the text and totally integrated with it; the reader is, thus, strongly encouraged to solve the problems in the appropriate context (and compare his or her answers to those given at the end of the book), but it is not compulsory to do that in order to continue the reading and learning process.

The bibliography includes 85 papers on the teaching of Chemistry and Physics (73 of which are from the *Journal of Chemical Education*). Accordingly, the book reflects the up-dated views conveyed by the numerous publications especially in the *Journal of Chemical Education* related to the subject. It was the critical use of all this information together with his own experience and discussions with colleagues that enabled the author to avoid the enigma and the misinterpretations frequently encountered in introductions to quantum theory in many chemistry books and to produce a simple, yet correct, text. Examples of topics which deserved detailed attention because they are often objects of misinterpretation are the wave–particle duality and the Heisenberg principle, the question of the electronic energy in terms of the orbital contributions, the use (and misuse) of hybrid orbitals and localized descriptions of the chemical bonding and the widely used qualitative models of molecular geometry.

Here is a brief description of the contents. Setting the stage for the main chapters, two short introductory chapters are devoted to the fundamentals of quantum mechanics: energy quantization; wave–particle duality in relation to observations, wavefunctions and the indeterminacy principle; the main operators and their properties, the Schrödinger equation for simple systems and, briefly, spin and relativity. Atomic and molecular orbitals for one-electron systems are introduced in Chapters 3 and 4; molecular orbitals are approximated by linear combinations of atomic orbitals; total electron angular momentum is analyzed. The adaptation of the orbital concept to many-electron atoms is dealt with in Chapter 5: Hartree–Fock orbitals, Kohn–Sham orbitals and electron correlation; electron configurations, terms and levels for open-shell atoms are discussed; some attention is also paid to relativistic effects. Molecular orbitals and chemical bonding in diatomic and polyatomic molecules are presented in Chapters 6 and 7, respectively, taking advantage of symmetry; electronegativity scales are compared; an introduction is made to the quantitative determination of molecular orbitals for any molecule, at different levels of approximation. Chapter 8 establishes the bridge between molecular orbitals, which are by definition delocalized functions, and the classical language used by chemists in their discussion of bonding and other electronic properties, namely the concepts of electron pair bonding and structural formulae: an introduction to Bader's theory of atoms in molecules, a discussion of canonical molecular orbitals and localized (or quasi-localized) functions, an introduction to electron pairing in valence-bond theory, a critical analysis of the VSEPR method in the interpretation of molecular geometry, a thorough discussion of the uses and misuses of hybrid orbitals. The theme of structural formulae is resumed in

Chapter 10, in relation to classical bond orders and bond energies. In the meantime, Chapter 9 deals with the $\sigma-\pi$ molecular orbital separation, conjugated systems, non-localizable π molecular orbitals and resonance. In Chapter 11 a brief extension of molecular orbital theory is made to include three categories of systems: fullerenes, transition metal complexes, solid aggregates (and band theory). Finally, Chapter 12 mainly illustrates the direct relations between orbitals and chemical reactivity and between orbitals and spectroscopy, with emphasis on electronic transitions and on spectral parameters in NMR spectroscopy.

It is hoped that the present book, which is of primary interest to undergraduate chemistry students, will provide, in an economic way, the basic preparation in atomic and molecular structure and chemical bonding theory needed for the interpretation of organic chemistry, inorganic chemistry and spectroscopy.

Victor M. S. Gil
Coimbra (Portugal), September 1999

Acknowledgements

My first thoughts of gratitude go to (the late) Professor F. Pinto-Coelho of the University of Coimbra (Portugal), for teaching me about orbitals when I was an undergraduate student, and to Professor John N. Murrell of the University of Sussex, (U.K.), for making me use the concept in a critical manner when I was a Ph.D student of his at the University of Sheffield (U.K.) and for his guidance and collaboration ever since.

Thanks are also due to many of my colleagues and students who directly or indirectly contributed to my education in this field. The more recent help of many is gratefully acknowledged, especially the theoretical physicists Helena Caldeira and Carlos Fiolhais (for discussions on the wave–particle duality and the Heisenberg principle and on density-functional theory, respectively), the Ph.D. students Jorge Trindade and Sérgio Rodrigues (for help with the numerical calculations and the graphical displays of molecular orbitals), the chemist Paula Catarro (for the molecular orbital diagrams), the chemists Christoffer Brett, Hugh Burrows, Brian Goodfellow and Ana Gil (for criticizing the text and correcting the language mistakes). Thanks are also extended to the reviewers of Cambridge University Press.

Finally, I am thankful to those periodicals and authors who permitted the reproduction of figures and to the Gulbenkian Foundation (Lisbon) for allowing the use of much of the material which I have included in a Portuguese book previously edited by them.

1

Energy, probability and electrons

1.1 Energy quantization

The interpretation of chemical phenomena, and hence the development of chemistry, owes a great deal to two fundamental concepts: *energy* and *probability*. The scientific idea of energy only emerged during the nineteenth century, whereas the notion of probability is at least two centuries older. In modern chemistry, there is hardly any field which does not depend upon one or both of these basic concepts, quite often coupled to each other. Important examples where energy and probability converge simultaneously are the second law of thermodynamics and entropy (in the field of chemical transformations), and quantum chemistry and orbitals (in the field of structure of atoms and their groupings). Orbitals are the main subject of this book. In so far as the orbital concept is essential to the study of the structure of matter it also lies in the realm of physical properties – such as electric, magnetic, spectroscopic and related properties – and of chemical behaviour of substances, from both the point of view of kinetics and thermodynamics. As we will briefly see below, the links between the fields from which the two previous examples were drawn have long been established; especially the historical role played by thermodynamics in the foundations of our knowledge about matter and light is recalled.

Orbitals are mathematical functions of the coordinates of each electron in atoms, molecules and other atomic aggregates; we will see that they carry the dimensions $[\text{length}]^{-3/2}$. They contain information on the *probability distribution* for each electron in space and correspond to certain electronic *energy values*. There are two most significant non-classical features which concern the energy of bound electrons. One is that not all values of energy are allowed: *energy quantization*. The other is the so-called *exchange energy* which is a consequence of the indistinguishability of electrons. In addition,

1

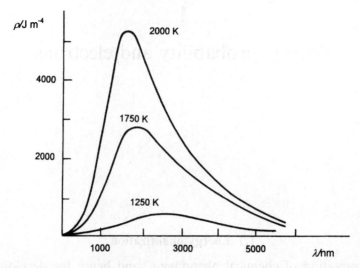

Fig. 1.1 Distribution of the energy output for a black-body (the ideal case of a body that neither reflects nor transmits light; it just absorbs or emits) at different temperatures as a function of wavelength.

our knowledge of the position and momentum of electrons is affected by the *Heisenberg indeterminacy principle*.

The concept of energy quantization has its roots in two kinds of experimental data gathered during the second half of the nineteenth century, both related to light: (a) the discontinuous emission spectra of gaseous elements; and (b) the distribution of the light intensity emitted by heated bodies as a function of wavelength for various temperatures (the so-called black-body) (Fig. 1.1).

The discrete wavelengths λ of the emission spectra were found by the end of the nineteenth century to be reproduced by a simple empirical formula, the Balmer–Rydberg equation:

$$1/\lambda = \mathcal{R}(1/n_1^2 - 1/n_2^2) \tag{1.1}$$

where $\mathcal{R} = 1.097 \times 10^5 \text{ cm}^{-1}$ is a constant (the Rydberg constant) and n_1 and n_2 ($n_1 < n_2$) are integer numbers. At about the same time (1890), the German physicist Max Planck (1858–1947, 1918 Nobel laureate in Physics) found another empirical formula which reproduces the density of radiating energy due to a black-body, as a function of frequency (ν) and temperature (T):

$$\rho(\nu, T) = A\nu^3/(e^{B\nu/T} - 1) \tag{1.2}$$

where A and B are constants.

The strange results summarized in Eq. (1.1) were in apparent conflict with thermodynamic data, especially with the measured heat capacities of gaseous elements. Meanwhile, Planck, taking advantage of his experience in thermodynamics, sought an explanation of Eq. (1.2) on the basis of the statistical interpretation of entropy that had been developed in 1896 by another Austrian physicist, Ludwig E. Boltzmann (1844–1906). Following previous ideas, Planck began by considering the atoms at the surface of a black-body as oscillators capable of absorbing and emitting light. In his studies and as a mathematical convenience, Boltzmann had often begun by considering the energy in small discrete amounts and only at the end of his analysis allowed this discontinuity to be removed. Planck used a similar treatment and considered discrete values $0, h\nu, 2h\nu, 3h\nu, \ldots$ (more or less populated according to the temperature) for the energies of the atomic oscillators, ν being the frequency of the oscillation and h a constant that, at the final stage, should be set equal to zero to allow all vibrational energies to be taken into account. It was then found by Planck that agreement with experiment (that is, Eq. (1.2)) could only be reached if h remained non-zero and was set equal to 6.63×10^{-34} J s. The theoretically derived counterpart of Eq. (1.2) is

$$\rho(\nu, T) = (8\pi h\nu^3/c^3)/(e^{h\nu/kT} - 1) \qquad (1.3)$$

where c is the speed of light in vacuum and k is the Boltzmann constant. The constant h was later, very properly, named the Planck's constant. Meanwhile, $h\nu$, the difference between successive energy values, was called a *quantum* of energy for the electromagnetic oscillators. The concept of energy quantization was born (with important contributions from the work of other physicists, mainly Franck and Hertz). Although the energy difference $h\nu$ between successive energy values was later found to be in agreement with the quantum-mechanical treatment of diatomic molecules as harmonic oscillators, the minimum value, zero, considered by Planck should be replaced by $h\nu/2$, the so-called zero-point energy. For an harmonic oscillator, the allowed vibrational energy values are $h\nu/2, 3h\nu/2, 5h\nu/2, \ldots$.

In 1905, Albert Einstein (1879–1955, 1921 Nobel laureate in Physics), born in Germany and later a naturalized American, having overcome some initial disbelief concerning the energy quantization of oscillators, extended this idea to electromagnetic radiation itself. Radiation carries energy in discrete amounts – *quanta* – each quantum or 'particle' of radiating energy being

$$E = h\nu \qquad (1.4)$$

where ν is now the frequency of the electromagnetic radiation. It has been recognized that Einstein became much more enthusiastic about Planck's theory than the author himself, who had great difficulty in discarding the principles of mechanics – now called classical mechanics – on which he had been brought up. The 'particles' of radiating energy were some years later named *photons* by the American chemist Gilbert N. Lewis (1875–1946). Thus, the energy of η photons corresponding to a given frequency ν is

$$E = \eta h\nu. \tag{1.5}$$

The photon model explained more than the black-body radiation. One of the first great achievements of the new theory of light was the interpretation of the *photoelectric effect*. The difference between the energy $h\nu$ of the incident photon and the minimum energy I necessary to remove an electron from the structure it belongs to (ionization energy, in the case of gaseous samples; work function, in the case of condensed phases) appears as kinetic energy $mv^2/2$ of the ejected electron:

$$h\nu = I + mv^2/2. \tag{1.6}$$

Thus, electron ejection requires a minimum frequency ν, it being irrelevant to have an intense source of light (many photons per unit of area) if the frequency is less than that minimum. However, the more intense the beam of light of appropriate frequency the greater the number of electrons ejected, by a one-to-one photon–electron interaction.

The proposal of Einstein was not accepted easily. Even as late as 1913, Planck himself, when joining other distinguished German physicists in recommending Einstein's appointment to the Prussian Academy of Sciences, would write

... That he may sometimes have missed the target in his speculations, as, for example, in his hypothesis of light quanta, cannot really be held against him, for it is not possible to introduce fundamentally new ideas, even in the most exact sciences, without occasionally taking a risk.

The idea of energy quantization was brought into chemistry with the application of quantum theory to the electronic structure of atoms in 1913 by the Danish physicist Niels Bohr (1885–1962, 1922 Nobel laureate in Physics). At the time, Bohr was working in the laboratory of the New Zealand physicist Ernest Rutherford (1871–1937, 1909 Nobel laureate in Chemistry) in England, a short time after the nuclear structure for the atom had been established by Rutherford and his co-workers. Classical electromagnetic theory predicted that the electrons around the nucleus, because they are being accelerated (centripetal acceleration), should radiate energy and thus should gradually approach the nucleus, emitting energy in a continuous manner, until the final collapse into the nucleus. However, it was known that atoms are not only stable entities but also, when previously excited, they radiate energy in discrete amounts. Then, inspired by the quantum theory of Planck, Bohr was forced to admit that the electrons in atoms exist in *stationary energy states*, with a well-defined energy, which they can only leave by either absorption or emission of certain discrete values of energy.

In such states, the electrons (charge $-e$ and mass m_e) would have circular orbits (radius r) around the nucleus, undergoing transitions between orbits through either absorption or emission of energy. In particular, for the hydrogen atom, the total energy of the electron in circular orbit (the kinetic energy in the rest frame of the nucleus, $m_e v^2/2$, plus the potential energy associated with the electron–nucleus attraction, $-\ell e^2/r$) could only have certain values. In order to obtain these values, Bohr began by assuming, as was done by Planck for the radiation of the black-body, that the energy emitted by the excited H atom was given by

$$E = n'h\nu' \tag{1.7}$$

with n' a positive integer and ν' a frequency characteristic of the motion of the electron around the nucleus. The crucial problem was then to relate ν' with the angular velocity of the electron circular motion. Bohr made some conjectures that, however, led to an impasse. When, almost by chance, the Balmer–Rydberg equation (1.1) was made known to him, the solution became clear and the following formula could be established for the possible values of energy of the electron in the H atom:

$$E_n = -(2\pi^2 e^4 m_e)/h^2 n^2 \tag{1.8}$$

with $n = 1, 2, 3, \ldots$ a *quantum number*.

This energy quantization would imply, too, quantization of the angular momentum of the electron $m_e vr$:

$$m_e vr = nh/2\pi. \tag{1.9}$$

Contrary to what is often found in books, Eq. (1.9) was not the starting point for the energy quantization expression (1.8), but the other way round. (For discussion of this point and the presentation of the Bohr model, see, for example, refs. 1 and 2.)

However, the theory presented several limitations and could not explain the spectra of atoms other than monoelectronic atoms. Bohr himself would recognize the need for a new theory and, indeed, would contribute to it. In particular, it is noted that, although Eq. (1.8) is reproduced by quantum mechanics, the angular momentum of the electron is not given by Eq. (1.9); in particular, the minimum value possible is zero and not $h/2\pi$ as required by Eq. (1.9).

1.2 The wave–particle duality, observations and probability

The photon theory of light and Eq. (1.5) received additional confirmation when, in 1923, the American physicist Arthur H. Compton (1892–1962, 1927 Nobel laureate in Physics) discovered the effect that would bear his name: *the Compton effect*. When X-rays interact with electrons, the scattered radiation, after transferring some energy to stationary electrons (which are accelerated by the electric field of the radiation), has a slightly higher wavelength. In contrast with classical physics, this increase $\Delta\lambda$ depends only on the angle θ through which the radiation is deflected, being independent of the initial wavelength:

$$\Delta\lambda = \lambda' - \lambda = \lambda_c(1 - \cos\theta). \tag{1.10}$$

The constant $\lambda_c = 2.425$ pm is called the *Compton wavelength* of the electron. The wavelength shift given by Eq. (1.10) can be easily reproduced theoretically if the interaction between the radiation and the electron is considered as a collision between two particles in which the energy and the linear momentum are conserved (conservation of momentum in the incident direction and in the direction perpendicular to it). These particles are a photon of energy $h\nu$ and linear momentum $p = h\nu/c = h/\lambda$ and a stationary electron of mass m_e which acquires velocity v (Fig. 1.2). It is then found

$$\lambda_c = h/m_e c = 2.425 \text{ pm}. \tag{1.11}$$

Equation (1.5) establishes a bridge between a description of light as an (electromagnetic) wave of frequency ν and as a beam of η energy particles. If phenomena related to time averages, such as diffraction and interference, can be easily interpreted in terms of waves, other phenomena, involving a one-to-one relation such as the photoelectric and the Compton effects, require a description based on corpuscular attributes. This wave–particle duality reflects the use of one or the other description depending on the experiment performed, while no experiment exists which exhibits both aspects of the duality simultaneously.

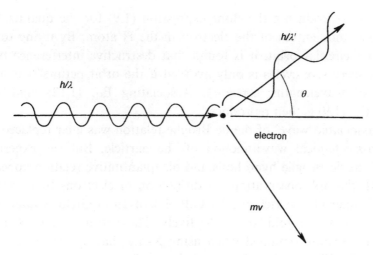

Fig. 1.2 Conservation of linear momentum in the Compton effect.

The form $p = h\nu/c = h/\lambda$ for the linear momentum of a photon can be obtained by combining the expression for the energy of a photon $E = h\nu$ and the expression $E = m_{ph}c^2$ which defines the relativistic mass of the photon, m_{ph}, provided that the linear momentum of a photon is made equal to $m_{ph}c$ by analogy with the classical expression mv for the linear momentum of a particle.

Strongly influenced by the interpretation of the Compton effect, the French physicist Louis Victor, Prince de Broglie (1892–1987, 1929 Nobel laureate in Physics), suggested in his doctoral thesis in 1924 that the wave–particle duality for photons could be extended to any particle of momentum $p = mv$ which, *somehow*, would then have a wavelength – the *de Broglie wavelength* – associated with it and given by

$$\lambda = h/p. \tag{1.12}$$

Indeed, considering that photons are rather peculiar particles in that they have zero rest mass and can exist only when travelling at the speed of light, it seems reasonable that the wave associated with the motion of any particle should become more and more apparent as the mass decreases, rather than the wave coming into existence suddenly when the rest mass vanishes.

This was a revolutionary and quite nebulous suggestion, not easily accepted at the time. If it was not for the intervention of Einstein, excited with the proposal, Louis de Broglie would most likely have failed his doctoral examination. Louis de Broglie found support for his hypothesis and attempted to clarify the 'wave characteristics' of a moving electron by successfully reproducing the Bohr expression (1.9) for the quantization of the angular momentum of the electron in the H atom. By trying to adjust waves to a circular orbit, it is found that destructive interference between waves in successive cycles is only avoided if the orbit perimeter is a whole number of wavelengths: $2\pi r = n\lambda$. Associating Eq. (1.12), and putting $p = m_e v$, Eq. (1.9) is then reproduced.

The 'associated wave' of the de Broglie relation was later replaced by the quantum-mechanical wavefunction of the particle. But the experimental proof of the de Broglie hypothesis and his quantitative relation appeared in 1927 with the first observations of diffraction of electrons by C. Davisson and L. Germer in America and by G.P. Thomson in Great Britain, using a nickel crystal and a gold foil, respectively. These results were found to be analogous to those obtained when using X-rays having wavelengths equal to the de Broglie wavelengths for electrons. It is interesting to note that, whereas J.J. Thomson showed at the end of the nineteenth century that the electron 'is a particle', and received the 1906 Nobel prize in Physics mainly for that, his son G.P. Thomson showed that it 'is a wave', and got the same prize in 1937 (shared with Davisson). In 1932, the occurrence of diffraction of helium atoms and hydrogen molecules by crystals was also found to be in agreement with the de Broglie relation.

The interpretation given by de Broglie for the quantization of the angular momentum of the electron of H, in the Bohr model, assumes *in some sense* that the electron can interfere with itself. Although any diffraction experiment always presupposes a large number of particles, it is not necessary that all the particles are considered at the same time; diffraction can still be obtained with a sequence of a large number of particles, one at a time. For example, a diffraction pattern is gradually built up as photons or electrons pass through two slits which are separated by a distance of the order of the wavelength of the particles (Fig. 1.3). In this sense, it can be said that each particle interferes with itself (refs. 3–5).

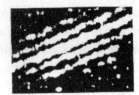

Fig. 1.3 Gradual genesis of an electron interference pattern in a double-slit experiment, with electrons reaching the detector one at a time. (Adapted with permission from ref. 5.)

This result clearly points to the impossibility of considering that the particle passes through one slit and not through the other; it seems 'to pass through both'. In other words, we can only say that it has 50% probability of passing through one slit and 50% of passing through the other. A bridge between the *wave characteristics* of small particles and *probability* is beginning to arise. Simultaneously, the notion of trajectory – from one slit to the screen or detector – no longer applies.

In contrast to this, for macroscopic particles, it is possible to identify the slit through which they pass; trajectories can then be defined and no interference occurs. The passages through each slit are then independent events and the probability of particles striking a given point on the screen, in a certain time interval, is just the sum of the probabilities corresponding to passage through slit 1 and through slit 2.

The interference of microscopic particles leads to a diffraction pattern with deviations with respect to the mere sum of the individual probabilities. The two events are no longer independent. If we wish to state in advance where the next particle will appear, we are unable to do so. The best we can do is to say that the next particle is more likely to strike in one area than another. *A limit to our knowledge*, associated with the wave–matter duality, becomes apparent. In the double-slit experiment, we may know the momentum of each particle but we do not know anything about the way the particles traverse the slits. Alternatively, we could think of an experiment that would enable us to decide through which slit the particle has passed, but then the experiment would be substantially different and the particles would arrive at the screen with different distributions. In particular, the two slits would become distinguishable and independent events would occur. No interference would be detected, that is, the wave nature of the particle would be absent. In such an experiment, in order to obtain information about the particle position just beyond the slits, we would change its momentum in an unknown way. Indeed, recent experiments have shown that interference can be made to disappear and reappear in a 'quantum eraser' (ref. 6 and references therein).

This discussion extends to electrons, and small particles in general, what we have already found for photons, that is, the appearance of electrons behaving as classical particles or as a wave depends on the experiment being performed. On the other hand, it illustrates a basic feature of microsystems: there is an unavoidable and uncontrollable interaction between the observer and the observed. No matter how cleverly one devises an experiment, there is always some disturbance involved in any measurement and such a disturbance is intrinsically indeterminate (see also Section 1.3).

However, the situation concerning the limits to our knowledge depicted above is not as desperate as it would seem. The fact that we can say that the next particle in a double-slit diffraction experiment is more likely to strike in one area of the screen than another points to the existence of a certain 'determinism' in the statistical result for a large number of identical experiments. We need to consider not each individual microsystem but a *statistical ensemble*, that is a great number of non-interacting replicas of a given microsystem which enable a large number of identical experiments to be performed; or, alternatively, we need to repeat the same experiment with the same microsystem, always in the same conditions. For example, one mole of H atoms corresponding to 6.022×10^{23} atoms all in the same electronic state is an ensemble for measurements about the electron. This characterization of a collection of microsystems in the same conditions is called *state preparation*.

Each individual measurement of any physical quantity yields a value A. But, independently of any possible observation errors associated with imperfect experimental measurements, the outcomes of identical measurements in identically prepared microsystems are not necessarily the same. The results fluctuate around a central value. It is this collection or spectrum of values that characterizes the *observable* A for the ensemble. The fraction of the total number of microsystems leading to a given A value yields the probability of another identical measurement producing that result. Two parameters can be defined: the *mean value* (later to be called the 'expected value') and the *indeterminacy* (also called uncertainty by some authors). The mean value $\langle A \rangle$ is the weighted average of the different results considering the frequency of their occurrence. The indeterminacy ΔA is the standard deviation of the observable, which is defined as the square root of the dispersion. In turn, the dispersion of the results is the mean value of the squared deviations with respect to the mean $\langle A \rangle$. Thus,

$$\Delta A = \langle (A - \langle A \rangle)^2 \rangle^{1/2} \tag{1.13}$$

which is a statistically meaningful expression of precision or reproducibility of measurements. (For a discussion of precision and accuracy in measurements, see, for example, ref. 7.)

1.3 Wavefunctions and the indeterminacy principle

Much of what has been said earlier lies at the foundations of the new mechanics which began with the work of the German physicist Max Born

System

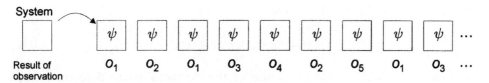

Result of
observation

Fig. 1.4 A quantum ensemble is constructed by replicating a microsystem many times, all members prepared in the same way so that they are in the same state (represented by ψ).

(1882–1970, 1954 Nobel laureate in Physics) in 1924. In quantum mechanics, two ensembles which show the same distributions for all the observables are said to be in the same *state*. Although this notion is being introduced for statistical ensembles, it can also be applied to each individual microsystem (see, for example, ref. 8), because all the members of the ensemble are identical, non-interacting and identically prepared (Fig. 1.4). Each state is described by a *state function*, ψ (see, for example, ref. 3). This state function should contain the *information* about the probability of each outcome of the measurement of any observable of the ensemble. The wave nature of matter, for example the interference phenomena observed with small particles, requires that such state functions can be superposed just like ordinary waves. Thus, they are also called *wavefunctions* and act as probability amplitude functions.

In particle diffraction experiments, the probability distribution $P(x, t)$ of the particles, electrons say, in the detector (x coordinate) should be reproduced from a wavefunction $\psi(x, t)$. At each value of time, the function $P(x, t)$ is large at values of x where an electron is likely to be found and small at values of x where an electron is unlikely to be found. Using $\psi(x, t)$ as a probability amplitude function which, to be more general, is allowed to be a complex function, the relation is

$$P(x, t) \propto \psi^*(x, t)\psi(x, t) \tag{1.14}$$

where $\psi^*(x, t)$ is the complex conjugate of $\psi(x, t)$: if $\psi(x, t)$ is of the form $a + ib$, $\psi^*(x, t)$ is $a - ib$ and $P(x, t) = a^2 + b^2$, a necessarily positive quantity. The relation (1.14) resembles that between the probability of finding photons at a given point and the square of the electromagnetic field at that point.

In a double-slit experiment, two functions can be defined, $\psi_1(x, t)$ and $\psi_2(x, t)$, for the two possibilities of the electron behaviour, one corresponding to passage through slit 1, the other to slit 2. The appropriate wavefunction is the *superposition* of $\psi_1(x, t)$ and $\psi_2(x, t)$, that is,

$$\psi(x, t) = \psi_1(x, t) + \psi_2(x, t), \tag{1.15}$$

and the probability density function of the electrons at the detector is then given by

$$\begin{aligned}
\psi^*(x, t)\psi(x, t) &= [\psi_1^*(x, t) + \psi_2^*(x, t)][\psi_1(x, t) + \psi_2(x, t)] \\
&= \psi_1^*(x, t)\psi_1(x, t) + \psi_2^*(x, t)\psi_2(x, t) + \psi_1^*(x, t)\psi_2(x, t) \\
&\quad + \psi_2^*(x, t)\psi_1(x, t). \tag{1.16}
\end{aligned}$$

The last two terms in Eq. (1.16) represent interference. This can be made more clear if the complex functions are written in terms of their magnitudes and phases:

$$\psi_1(x, t) = |\psi_1(x, t)| \exp[i\alpha_1(x, t)]$$
$$\psi_2(x, t) = |\psi_2(x, t)| \exp[i\alpha_2(x, t)]. \tag{1.17}$$

It is easily found that substitution of Eq. (1.17) into Eq. (1.16) leads to:

$$\psi^*(x, t)\psi(x, t) = |\psi_1(x, t)|^2 + |\psi_2(x, t)|^2 + 2|\psi_1(x, t)| \, |\psi_2(x, t)| \cos(\alpha_1 - \alpha_2). \tag{1.18}$$

The interference term depends on the phase difference $\alpha_1 - \alpha_2$ of the probability amplitude functions.

We have mentioned in Section 1.2 that when enough information is available to determine which path is taken by the particle, the interference disappears, and it reappears when the path cannot be determined. The explanation for this is not that quantum waves refrain from interfering when observed too closely, it is just that information about the path is available when the quantities measured are not sensitive to the phase difference $\alpha_1 - \alpha_2$, but quantities that are sensitive to this phase difference can be measured when those that determine the path are not (ref. 6).

The example of the double-slit experiment already shows that the physical information contained in a state function is inherently probabilistic in nature. In the next chapter this feature will be further developed leading to the central concept of this book: *orbitals*. For example, the orbitals of the H atom are wavefunctions $\psi(x, y, z)$ that enable the electron probability density to be known: $\psi^*(x, y, z)\psi(x, y, z)$. The idea of a trajectory (or orbit) is replaced by the idea of a probability distribution.

The above interpretation of the wavefunction, linking a mathematical function with a physical state, is due to Born and provides the basis of the most widely held view of quantum mechanics: the *Copenhagen interpretation*. Helped by this interpretation, other scientists rapidly developed the theoretical machinery for extracting information about position, momentum and other observables from wavefunctions, during the years 1925–1927. In 1925, the German physicist Werner Heisenberg (1901–1976, 1932 Nobel laureate in Physics), disciple of Max Born, represented the physical quantities by complex numbers in a mathematical treatment that Born clarified by establishing a relation with the properties of matrices. This work was in part developed in collaboration with the German physicist Pascual Jordan whom Born met, by accident, during a train journey.

In the same year, the English mathematician (of Swiss father) Paul Dirac (1902–1984, 1933 Nobel laureate in Physics) developed a new formalism strongly related to classical mechanics through a collection of postulates, which is equivalent to that of Heisenberg but more easy to understand. Almost at the same time, the Austrian physicist Erwin Schrödinger (1887–1961, 1933 Nobel laureate in Physics, prize shared with Dirac) proposed a wave approach to quantum mechanics, inspired by the de Broglie theory which he had initially classified as nonsense. As we shall see in Chapter 2, the form of his wave equation is analogous to the equations of classical electromagnetic theory, the wavelength being given by the de Broglie relation. It is a differential equation, involving second-order derivatives of the state function of the system, ψ, one of the unknowns, and a second unknown which is the energy E. By introducing basic conditions which ψ must obey because of its relation with probability densities, the solution of the wave equation can lead to a discrete set of E values; once these values are known, the wavefunctions ψ associated with them are then obtained. In this way, energy quantization appears as a direct consequence of the physical interpretation of the wavefunction.

Initially, Schrödinger's theory was not well accepted by either Heisenberg or Dirac, but gained the support of Born. The situation changed when, in 1926, Schrödinger showed that his theory was equivalent to those of Heisenberg and Dirac. Beginning with the Heisenberg–Born equations involving matrices to represent physical quantities, Schrödinger showed that these could be represented by appropriate *operators* (see Chapter 2). In the particular case of the electron in the H atom, the operator corresponding to the observable energy is related to the Hamilton function of classical mechanics, modified previously by Dirac.

In 1927, Heisenberg identified *incompatible observables* whenever the commutative property of multiplication does not apply to the corresponding operators and showed that any attempt to simultaneously measure two incompatible physical quantities A and B necessarily introduces an imprecision in each observable. No matter how the experiment is designed, the results are inevitably indeterminate, and the indeterminacies ΔA and ΔB cannot be reduced to zero. Instead their product is always larger than a constant of the order of Planck's constant. The easiest practical way to recognize incompatible variables is to note that their dimensions multiply to joule second. For example, position x (y, or z) and momentum p_x (p_y, or p_z) are fundamentally incompatible observables, in the sense that knowing the precise value of one precludes knowing anything about the other. Thus,

$$\Delta x \, \Delta p_x \geq h/4\pi \qquad \Delta y \, \Delta p_y \geq h/4\pi \qquad \Delta z \, \Delta p_z \geq h/4\pi. \qquad (1.19)$$

This is known as the *Heisenberg indeterminacy principle* (also called uncertainty principle by some authors). It has to do with precision and not with accuracy. This situation has already been met in Section 1.2 when referring to the double-slit experiment.

On the scale of classical mechanics, the indeterminacies can, at least in principle, be made sufficiently small to be negligible. Then, the order in which we measure position and linear momentum is arbitrary. Irrespective of this order, we may have, for example, 2 m for the position x of a body (with respect to some origin) and $3 \, \text{kg m s}^{-1}$ for its momentum p_x which implies $xp_x = 2\,\text{m} \times 3\,\text{kg m s}^{-1} = p_x x = 3\,\text{kg m s}^{-1} \times 2\,\text{m} = 6\,\text{kg m}^2\,\text{s}^{-1} = 6\,\text{J s}$. However, for an ensemble of microsystems, the indeterminacies are no longer simultaneously negligible. In such a case, we need matrices or operators to represent those observables, and $xp_x \neq p_x x$. Heisenberg showed that $xp_x - p_x x = ih/2\pi$ (see Chapter 2). This is equivalent to saying that the order of preparation of the ensemble for adequate measurements is not irrelevant, as shown below.

Let us assume that we begin by preparing a state ψ so that identical measurements of the momentum p_x in the various members of the ensemble have an arbitrarily small dispersion. Then, if, in a separate experiment we prepare the same state ψ and make identical individual measurements of the position x, the result is a huge dispersion of the x values. If, on the other hand, we begin by forcing (preparing) the ensemble to a state ψ' in order to yield an arbitrarily small indeterminacy of position, then there will be a huge indeterminacy in the momentum values obtained through identical measurements of the various microsystems in state ψ' (Fig. 1.5). The impossibility of reducing both indeterminacies to an arbitrarily small value is intimately related to the fact that ψ is different from ψ'. In Chapter 2, we will say that no 'eigenfunction' for the position operator (corresponding to a

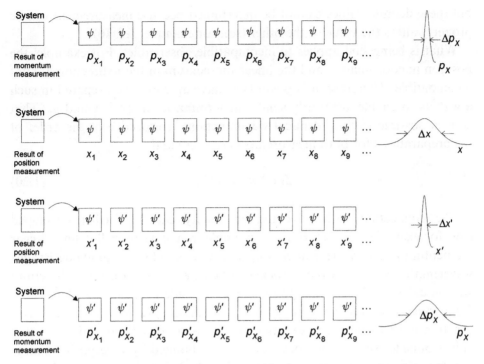

Fig. 1.5 Identical preparation of an ensemble for measurements of two incompatible observables A and B, so that one of them is affected by an arbitrarily small indeterminacy; the indeterminacy in the other is necessarily huge.

dispersion-free distribution of results) is an eigenfunction for the momentum operator, and vice versa.

It should be noted that, since we are talking about measurements in separate experiments, strictly speaking it is not the measurement of x (or p_x) that directly affects the measurement of p_x (or x); it is the preparation process of the ensemble for the two measurements which, being the same, relates the two measurements. Thus, the usual statement *the impossibility of knowing simultaneously with accuracy (rather, precision) both position and momentum of a particle* should be read as *the impossibility of preparing a state for which both position and momentum can be determined with arbitrarily small indeterminacies* (see, for example, ref. 9).

The limitation reflected in the Heisenberg principle is *implicit in Nature*. It has nothing to do with a particular apparatus or with the imperfections of experimental techniques. A particle *cannot* simultaneously have a precise value of x and a precise value of p_x. This is different from assuming that each particle really does have definite values for both position and momentum, but these definite values cannot be determined because measurement of one property alters the value of the other (see, for example, ref. 10).

What is being said applies to incompatible observables. For example, the position in coordinate x and the linear momentum in the y direction are not incompatible. Therefore, it is possible to have an ensemble prepared in such a way as to enable arbitrarily small indeterminacies in both x and p_y. That is, it is possible to find identical wavefunctions irrespective of the order of the preparations for both measurements ($\psi = \psi'$) and

$$\Delta x \, \Delta p_y \leq h/4\pi. \tag{1.20}$$

The connection with wave–matter duality can be further developed (see, for example, ref. 11). For an ensemble in a state of well-defined linear momentum p (along the x direction, say) we can define a unique de Broglie wavelength $\lambda = h/p$ and the particle can be at any point (in the x direction) with equal probability (this point will be considered again in Chapter 2). Conversely, if we know the position of the particle precisely, then the wavefunction must assume a large value at the point considered and very small values outside. Since such a function can be obtained by superposing a large number of waves of various wavelengths which interfere constructively at the point and destructively otherwise, the existence of various λ values means that all information about the linear momentum is lost (Fig. 1.6).

Obviously, if we cannot specify the position and the momentum of a particle, at any instant, we cannot maintain the concept of trajectory for small particles. This idea involves statements about the past, the present and the future of the system, and only the present is accessible to experimental determination and, even so, with limitations. According to Heisenberg, it

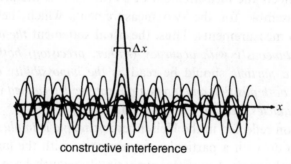

constructive interference

Fig. 1.6 A small indeterminacy in x as the result of appropriate superposition of waves having a variety of wavelengths: information about linear momentum $p = h/\lambda$ is entirely lost (see, for example, ref. 12).

makes no sense to ascribe properties to a system that cannot, at least in principle, be subjected to experimental determination. Properties of macro-systems exist by themselves, independently of being measured, and then we can speak both of precision and accuracy of measurements. This is not the case with microsystems. Here, properties are latent until they are the object of observation and measurement (at least in principle), and precision rules.

To conclude this introduction to quantum mechanics, it is interesting to note the omnipresence and the agglutinating role of Planck's constant. Indeed, if it was set equal to zero, all the construction which began with black-body radiation and the quantization of radiation energy, followed by the wave–matter duality and the Heisenberg principle ... would fall down. In addition, the intrinsic angular momentum (spin) of some particles, including the electron, would be forced to be zero, with many consequences at the theoretical and practical levels.

2

An introduction to the dynamics of microsystems

2.1 Operators and observables

When dealing with microphysical systems, the classical notion of trajectory for a moving particle must be abandoned, because, in contrast with Newtonian mechanics, we can no longer have a well-defined position and momentum at a given time for such a particle. Instead, we can only describe the probability of the particle being at a certain position in space, or the probability of it having a certain momentum. In quantum mechanics, trajectories are replaced by a spatial distribution; associated with any particle is a *wavefunction* ψ whose absolute square is proportional to the probability density for finding the particle. Indeed, the wavefunction (a state function) contains the information about the probability of each outcome of any observation. As we saw in Chapter 1, this presupposes that identical observations/measurements are made in a great number of non-interacting replicas of a given microsystem in a certain state represented by ψ (an *ensemble*) or, alternatively, that the same experiment is repeated a great number of times with the same microsystem in identical conditions. Because all members of the ensemble are identical, non-interacting and identically prepared, we shall often use the word system and may occasionally refer explicitly to the corresponding statistical ensemble.

In order to predict the outcome of an observation of a particular property for a given state of the system, it is necessary to perform appropriate operations such as multiplications or differentiations on the wavefunction. In particular, a measurable physical quantity or *observable* has associated with it a quantum-mechanical *operator* so that

$$(\text{operator})(\text{wavefunction}) = (\text{value of observable})(\text{wavefunction})$$

or, using the corresponding symbols,

$$\hat{O}\psi = o\psi. \tag{2.1}$$

The numerical value o of the observable is called the *eigenvalue* of operator $\hat{O}$ with respect to the state function ψ. For an operator to correspond to an observable, that is to produce an eigenvalue, it is necessary that the result of the operation on ψ is a numerical value times the wavefunction. The wavefunction is then an *eigenfunction* of the operator.

Usually, the problem is to find simultaneously ψ and the values o that satisfy the eigenvalue equation (2.1), the form of the operator having been previously established. An operator is a symbol telling us to carry out a certain mathematical operation on anything following it. An example is the differential operator $\hat{O} = d/dx$. It is easily found that $\psi = x^2$ is not an eigenfunction of this operator

$$dx^2/dx = 2x \neq ox^2 \tag{2.2}$$

whereas $\psi = e^{2x}$ is an eigenfunction of the same operator

$$d(e^{2x})/dx = 2(e^{2x}) \tag{2.3}$$

with $o = 2$.

Problem 2.1 Show that neither $\psi = \sin x$ nor $\psi = \sin x + \cos x$ are eigenfunctions of d/dx, whereas $\psi = \cos x - i \sin x = e^{-ix}$ is an eigenfunction with $o = -i$.

The wavefunctions can be real or complex functions. In the case of a complex wavefunction of the coordinates x, y, z of a particle, the probability of finding the particle at an element of volume dv around an arbitrary point is proportional to

$$\psi^*(x,y,z)\psi(x,y,z)\,dv \tag{2.4}$$

where $\psi^*(x,y,z)$ is the complex conjugate of $\psi(x,y,z)$. If the wavefunction is *normalized*, that is if $\psi(x,y,z)$ is affected by a numerical factor so that the summation of all elementary probabilities is unity

$$\int \psi^*(x,y,z)\psi(x,y,z)\,dv = 1 \tag{2.5}$$

then expression (2.4) represents the actual probability and $\psi^*(x,y,z)\,\psi(x,y,z)$ is the *probability density*.

An operator is classified as a *linear operator* if

$$\hat{O}(c\psi) = c\hat{O}\psi \tag{2.6}$$

where c is a real or complex number.

---

* The answers to the problems are given in the chapter beginning on p. 290.

Problem 2.2   Show that $\hat{O} = \mathrm{d}/\mathrm{d}x$ is a linear operator but $\hat{O} = \log$ or $\hat{O} = \sqrt{}$ are not.

If ψ_1 and ψ_2 are eigenfunctions of a linear operator, functions $c_1\psi_1 + c_2\psi_2$ are not eigenfunctions of that operator

$$\hat{O}(c_1\psi_1 + c_2\psi_2) = c_1\hat{O}\psi_1 + c_2\hat{O}\psi_2 = c_1 o_1\psi_1 + c_2 o_2\psi_2 \neq o(c_1\psi_1 + c_2\psi_2)$$

$$(2.7)$$

unless the eigenvalues o_1 and o_2 are equal: $o_1 = o_2 = o$. In this case, the eigenfunctions ψ_1 and ψ_2 are said to be *degenerate* with respect to the operator considered.

An operator is *hermitian* if

$$\int \psi_1^* \hat{O}\psi_2 \, \mathrm{d}\tau = \int \psi_2 \hat{O}^* \psi_1^* \, \mathrm{d}\tau \tag{2.8}$$

where the integration is over the entire range of the domain of the wavefunctions. Since the second term in Eq. (2.8) is equal to $(\int \psi_2^* \hat{O}\psi_1 \, \mathrm{d}\tau)^*$, the hermiticity criterion can be written in an abbreviated way called the Dirac bracket (bra-ket) notation:

$$\langle \psi_1 | \hat{O} | \psi_2 \rangle = \langle \psi_2 | \hat{O} | \psi_1 \rangle^*. \tag{2.9}$$

It can be proved that the eigenvalues of hermitian operators are real. Accordingly, observables are represented by hermitian operators, as that guarantees that the outcome of observations is real. Another property of those operators is that their non-degenerate eigenfunctions are *orthogonal*, that is

$$\int \psi_1^* \psi_2 \, \mathrm{d}\tau = 0$$

$$(2.10)$$

$$\langle \psi_1 | \psi_2 \rangle = 0.$$

Integrals of the form (2.10) are called *overlap integrals* between functions ψ_1 and ψ_2.

Problem 2.3   Relate the orthogonality of two functions with the scalar product of two vectors.

Problem 2.4   Using Eqs. (2.9) and (2.1) and taking $o_1 \neq o_2$ prove Eq. (2.10).

Wavefunctions that are normalized and orthogonal are said to be *ortho-normal*. Symbolically,

$$\langle \psi_m | \psi_n \rangle = \delta_{mn} \qquad \delta_{mn} = 1 \; (m = n) \qquad \delta_{mn} = 0 \; (m \neq n) \qquad (2.11)$$

where δ_{mn} is the Kronecker symbol.

A simple exercise shows that the eigenvalue o_n of an operator with respect to a certain eigenfunction ψ_n can be extracted from Eq. (2.1) by multiplying both members of the eigenvalue equation on the left by ψ_n^*, integrating and using Eq. (2.11):

$$\psi_n^* \hat{O} \psi_n = \psi_n^* o_n \psi_n = o_n \psi_n^* \psi_n$$

$$\langle \psi_n | \hat{O} | \psi_n \rangle = o_n \langle \psi_n | \psi_n \rangle = o_n. \qquad (2.12)$$

On the other hand, for two non-degenerate eigenfunctions ψ_m and ψ_n

$$\langle \psi_m | \hat{O} | \psi_n \rangle = o_n \langle \psi_m | \psi_n \rangle = 0. \qquad (2.13)$$

Therefore, we can say that the eigenfunctions of an operator $\hat{O}$ *diagonalize the matrix* $\hat{O}$, whose elements are

$$O_{nn} = \langle \psi_n | \hat{O} | \psi_n \rangle = o_n \qquad (2.14)$$

$$O_{mn} = \langle \psi_m | \hat{O} | \psi_n \rangle = 0. \qquad (2.15)$$

In the usual form, the matrix becomes

	ψ_1	ψ_2	ψ_3	ψ_4	$\cdots$
ψ_1	o_1	0	0	0	$\cdots$
ψ_2	0	o_2	0	0	
ψ_3	0	0	o_3	0	
ψ_4	0	0	0	o_4	

$$(2.16)$$

$\vdots$

2.2 Expectation values of observables

If the state of a system is an eigenstate of the operator corresponding to a given observable, all identical experiments give the same result which is the eigenvalue of the operator for such a state. An example is the determination of the energy of an atom or a molecule when it is in a stationary energy state; this is then an eigenstate of the energy operator of the system.

However, the state of a system is not an eigenstate of all operators. When the system wavefunction ψ is not an eigenfunction of the operator $\hat{A}$ representing a certain observable A, the outcomes of a series of identical experiments are not identical. The average of all outcomes (see Chapter 1) is the *expected value* or the *expectation value* of the operator with respect to the state under consideration:

$$\langle A \rangle = \langle \psi | \hat{A} | \psi \rangle. \tag{2.17}$$

This value is equivalent to a classical average of the physical observable A at a given time, where each contribution is weighted by the corresponding probability of occurrence given by the values of $\psi^* \psi$ (it is assumed that ψ is normalized). An example is the mean distance of the electron from the nucleus of atom H, for a given orbital:

$$\langle r_{1s} \rangle = \langle \psi_{1s} | r | \psi_{1s} \rangle$$
$$\langle r_{2p} \rangle = \langle \psi_{2p} | r | \psi_{2p} \rangle \tag{2.18}$$
$$\vdots$$

In the double-slit diffraction experiment considered in Chapter 1, the wavefunction is a superposition of two functions, one corresponding to the passage through slit 1 and the other through slit 2. According to the so-called *superposition principle* (see, for example, refs. 13 and 14), a given state function ψ of a system can be expressed in terms of the eigenfunctions φ_i (forming a complete, orthonormal set) of the operator $\hat{A}$ (i.e. $\hat{A}\varphi_i = a_i\varphi_i$) as follows:

$$\psi = c_1\varphi_1 + c_2\varphi_2 + c_3\varphi_3 + \cdots \tag{2.19}$$

The outcome of a given measurement of property A for the system in the state represented by ψ is unpredictable. It can be any of the eigenvalues of $\hat{A}$. In a large number of identical observations, the outcome is sometimes a_1, sometimes a_2, etc. according to a frequency which is proportional to c_i^2 (or $c_i^* c_i$ if the coefficients are complex numbers). Accordingly, the expectation value is the weighted average of the eigenvalues a_i:

$$\langle A \rangle = \langle \psi | \hat{A} | \psi \rangle = c_1^2 \langle \varphi_1 | \hat{A} | \varphi_1 \rangle + c_2^2 \langle \varphi_2 | \hat{A} | \varphi_2 \rangle + \cdots = c_1^2 a_1 + c_2^2 a_2 + \cdots \tag{2.20}$$

since all the cross integrals $\langle \varphi_i | \hat{A} | \varphi_j \rangle$ vanish.

Problem 2.5   Prove that all the cross integrals $\langle \varphi_i | \hat{A} | \varphi_j \rangle$ in the expansion (2.20) vanish.

Whenever an observable is required we encounter formulae like Eqs. (2.17), (2.18) and (2.20) in which the wavefunction appears as its square, or $\psi^*\psi$ for the general case of ψ being a complex function. For a one-particle system, $\rho = \psi^*(x,y,z)\psi(x,y,z)$ is the probability density function. For a two-electron system in a (time-independent) state described by the (spin-free) wavefunction $\psi(x_1,y_1,z_1,x_2,y_2,z_2)$, the density function is $\rho(1,2) = \psi^*(x_1,y_1,z_1,x_2,y_2,z_2)\psi(x_1,y_1,z_1,x_2,y_2,z_2)$ and $\rho(1,2)\,dv_1\,dv_2$ can be interpreted as the probability that electron 1 is in dv_1 (i.e. between x_1 and $x_1 + dx$, y_1 and $y_1 + dy$, and z_1 and $z_1 + dz$) and electron 2 is in dv_2. It is noted that the wavefunction $\psi(x_1,y_1,z_1,x_2,y_2,z_2)$ exists in a six-dimensional configuration space.

The electron density expression above is a particular case of a more general expression $\psi^*(x_1',y_1',z_1',x_2',y_2',z_2')\,\psi(x_1,y_1,z_1,x_2,y_2,z_2)$ where there are two sets of coordinate indices, one for each factor. This latter product may be thought of as an element of a matrix, with the continuously variable position coordinates playing the role of the row and column labels. This is called a *density matrix*. We will later refer to the *density-functional theory* of electronic structure as a collection of procedures used to solve many-electron problems in terms of the electron density, avoiding the many-electron wavefunction.

2.3 Commuting operators

Two operators $\hat{O}_1$ and $\hat{O}_2$ are said to commute if, for any wavefunction ψ,

$$\hat{O}_1\hat{O}_2\psi = \hat{O}_2\hat{O}_1\psi \tag{2.21}$$

which can be written

$$(\hat{O}_1\hat{O}_2 - \hat{O}_2\hat{O}_1)\psi = 0. \tag{2.22}$$

This requires that the *commutator* of $\hat{O}_1$ and $\hat{O}_2$, represented by $[\hat{O}_1, \hat{O}_2]$ vanishes:

$$[\hat{O}_1, \hat{O}_2] = \hat{O}_1\hat{O}_2 - \hat{O}_2\hat{O}_1 = 0. \tag{2.23}$$

An important property of commuting operators is that there is a set of functions that are eigenfunctions for both operators simultaneously (for the demonstration see, for example, ref. 13). This means that a particular measurement of property O_1 leading to an eigenvalue o_1 has also prepared the system to yield a particular value for the property O_2 so that a measurement of this latter property will yield the value o_2. Thus, the values of the two observables O_1 and O_2 represented by two commuting operators $\hat{O}_1$ and $\hat{O}_2$ may be simultaneously known and measured for a given state of the system, and the order of the two measurements is irrelevant (see Chapter 1).

If two operators do not commute with each other, then the observables they represent cannot be determined simultaneously with an arbitrarily small indeterminacy. These observables are said to be incompatible (also called complementary), and are the object of the indeterminacy principle introduced in Chapter 1. We will keep to this usual expression of the Heisenberg indeterminacy principle, although, as we found in Chapter 1, this means the impossibility of preparing a state for which two incompatible properties can be determined with arbitrarily small indeterminacies. The fact that some operators do not commute with each other represents one of the main differences between classical and quantum mechanics.

Problem 2.6   Find out whether the operators $\hat{O}_1 = \mathrm{d}/\mathrm{d}x$ and $\hat{O}_2 = ()^2$ commute with each other, that is check if $\hat{O}_1 \hat{O}_2 f(x)$ equals $\hat{O}_2 \hat{O}_1 f(x)$.

Problem 2.7   Determine the commutators of the operators x and $\mathrm{d}/\mathrm{d}x$, and y and $\mathrm{d}/\mathrm{d}x$.

One pair of incompatible observables is the position of a particle in a given coordinate (x) and its momentum in the same direction (p_x). The corresponding operators do not commute. This is in agreement with a non-classical form for the momentum operator $\hat{p}_x$. In fact, the classical commutator $x(mv_x) - (mv_x)x$ is zero.

2.4 Important operators

The linear momentum operator

According to the de Broglie equation, the linear momentum is related to a wavelength: $p = h/\lambda$. Therefore, a large linear momentum implies a wave of short wavelength (like a compressed spring) which, in turn, means a high gradient of the wavefunction. This points to a relation between p_x and $\partial \psi / \partial x$. On the other hand, the linear momentum operator must be hermitian, so as to have real eigenvalues. These considerations help the acceptance of the following form for the linear momentum operators

$$\hat{p}_x = -i\hbar \partial/\partial x$$
$$\hat{p}_y = -i\hbar \partial/\partial y \qquad\qquad\qquad (2.24)$$
$$\hat{p}_z = -i\hbar \partial/\partial z \qquad \hbar = h/2\pi = 1.055 \times 10^{-34}\,\mathrm{J\,s}$$

(the usual symbol $\hbar$, called h-bar, was introduced). We will see that such expressions lead to the Schrödinger equation which is, in turn, linked with the classical wave equation by the de Broglie relation.

Using Eq. (2.24), the commutator of operators x and $\hat{p}_x$ is easily established:

$$[x, \hat{p}_x]\psi = (x\hat{p}_x - \hat{p}_x x)\psi = x\hat{p}_x\psi - \hat{p}_x x\psi$$
$$= -i\hbar(x\partial\psi/\partial x - \psi - x\partial\psi/\partial x) = i\hbar\psi \qquad (2.25)$$
$$[x, \hat{p}_x] = i\hbar.$$

The energy operators

The form of the kinetic energy operator can be established from the form of the operator $\hat{p}$ by analogy with the classical expression for kinetic energy $p^2/2m$:

kinetic energy operator for the x direction: $-(\hbar^2/2m)\,\partial^2/\partial x^2$. $\qquad (2.26)$

Considering all the components, we have

$$-(\hbar^2/2m)(\partial^2/\partial x^2 + \partial^2/\partial y^2 + \partial^2/\partial z^2) = -(\hbar^2/2m)\nabla^2 \qquad (2.27)$$

where the operator ∇^2 is called the *laplacian* (see also ref. 15).

The second derivative of a function reflects its curvature. Therefore, high kinetic energy corresponds to high curvature of the wavefunction, which implies that the function goes rapidly from positive to negative amplitudes (see for example ref. 16). This means more *nodes*, hence a shorter wavelength. Indeed, according to the de Broglie relation, a shorter wavelength is equivalent to a larger momentum, hence to a higher kinetic energy. We will see later that orbitals having more nodes are associated with higher energy values.

As the operators associated with the spatial coordinates are just x, y, z (or r, θ, ϕ), so the operator for the potential energy – which is a function of those coordinates – is given by the classical expression. For the interaction of two electric charges q and q' at a distance r, the expression is

$V_e = kqq'/r$

$k = 1/4\pi\varepsilon_0 = 8.99 \times 10^9 \, \text{J m C}^{-2}$ (electromagnetic constant) $\qquad (2.28)$

$\varepsilon_0 = 8.85 \times 10^{-12} \, \text{J}^{-1} \, \text{m}^{-1} \, \text{C}^2$ (vacuum permittivity).

Finally, the *hamiltonian* which is the operator corresponding to the total energy E

$$\hat{H}\psi = E\psi \qquad (2.29)$$

is the sum of the kinetic energy operator and the potential energy operator:

$$\hat{H} = -(\hbar^2/2m)\nabla^2 + V. \qquad (2.30)$$

The energy and position operators do not commute, that is, no eigenfunctions for position can be eigenfunctions of $\hat{H}$. Therefore, if the energy is known the position will be indeterminate. We shall also see that, in contrast with classical mechanics, the energy of a constrained system cannot take any value.

The angular momentum operator

The classical expression for the magnitude of the angular momentum of a rotating particle having mass m and describing a circular path of radius r with speed v is mvr. Since it is also necessary to characterize the plane and the direction of the rotation, the *orbital angular momentum* is a vector equal to the cross product of the position vector $\mathbf{r}$ and the linear momentum $\mathbf{p}$:

$$\mathbf{l} = \mathbf{r} \times \mathbf{p}. \qquad (2.31)$$

The vector components are

$$
\begin{aligned}
l_x &= yp_z - zp_y \\
l_y &= zp_x - xp_z \\
l_z &= xp_y - yp_x
\end{aligned}
\qquad (2.32)
$$

and the magnitude is

$$l = (l_x^2 + l_y^2 + l_z^2)^{1/2}. \qquad (2.33)$$

Using the expressions for the linear momentum operators (2.24), the angular momentum operators are obtained:

$$
\begin{aligned}
\hat{l}_x &= -i\hbar(y\partial/\partial z - z\partial/\partial y) \\
\hat{l}_y &= -i\hbar(z\partial/\partial x - x\partial/\partial z) \\
\hat{l}_z &= -i\hbar(x\partial/\partial y - y\partial/\partial x).
\end{aligned}
\qquad (2.34)
$$

It is easily shown that these operators do not commute with each other. Therefore, it is not possible to find a set of functions which are simultaneous eigenfunctions of any two operators (2.34) and so it is not possible to determine with arbitrarily high precision and simultaneously any two components of the angular momentum. However, any of them commutes with the operator corresponding to the magnitude of the angular momentum vector: $\hat{l}^2$. Thus, the magnitude of the angular momentum and one of its components, l_z for example, are compatible observables and can be known simultaneously.

In addition, any of the operators $\hat{l}^2$ and $\hat{l}_z$ commute with the hamiltonian. Therefore, energy, the orbital angular momentum and its z component can be measured simultaneously. In contrast with classical mechanics where the magnitude and the orientation of the angular momentum vector can be continuously varied, according to quantum mechanics both the magnitude of the angular momentum and one of its components (the z component, say) are quantized; the projections on the other axes (x and y) cannot be specified. In Chapter 3, we will encounter wavefunctions for the electron in the H atom (orbitals) associated simultaneously with energy, orbital angular momentum and its z component, involving, respectively, the quantum numbers n, ℓ and m_ℓ.

We will also find that the angular momentum of the electron (and other particles) is not restricted to the orbital angular momentum. There is, in addition, an intrinsic angular momentum of the electron associated with its magnetic dipole moment. This is called the *spin angular momentum*. It is a non-classical property which is sometimes taken analogically as a rotation of the particle around one of its axes. The observable angular momentum of an electron is often called the total angular momentum because it can be assumed to be composed of orbital and spin contributions, summed as vectorial quantities.

2.5 The Schrödinger equation

Using the form of the energy operators, the Schrödinger equation can be established immediately. Thus, for the translational motion of a free particle (potential energy zero) along the x axis, we have

$$-(\hbar^2/2m)\,\mathrm{d}^2\psi/\mathrm{d}x^2 = E\psi \tag{2.35}$$

where E stands for kinetic energy.

Equation (2.35) can be related to the classical equation for a standing wave

$$d^2\Phi/dx^2 + (4\pi^2/\lambda^2)\Phi = 0 \qquad (2.36)$$

where Φ is the amplitude of the vibration given by

$$\Phi = C\sin(2\pi x/\lambda). \qquad (2.37)$$

Replacing E by $mv^2/2$ in Eq. (2.35) and rearranging both Eqs. (2.35) and (2.36), we get

$$-(\hbar^2/m^2v^2)\,d^2\psi/dx^2 = \psi \qquad (2.38)$$

and

$$-(\lambda^2/4\pi^2)\,d^2\Phi/dx^2 = \Phi. \qquad (2.39)$$

Equations (2.38) and (2.39) can be made identical by putting

$$\lambda^2/4\pi^2 = h^2/4\pi^2m^2v^2 \qquad (2.40)$$

that is

$$\lambda = h/mv \qquad (2.41)$$

the de Broglie relation.

Problem 2.8 Considering the solution $\psi = \sin kx$ of Eq. (2.35) with $k = (2mE)^{1/2}\hbar$, as a sine wave $\sin(2\pi x/\lambda)$ of wavelength $\lambda = 2\pi/k$, reproduce the de Broglie equation $p = h/\lambda$.

If Eq. (2.36) represents the oscillation of the electric field (or the magnetic field) of an electromagnetic radiation, then the light intensity at point x is proportional to the mean squared amplitude $\Phi^2(x)$. Since, in photonic terms, light intensity is proportional to the probability of finding photons at the point, per unit of time, there is proportionality between the photonic probability and $\Phi^2(x)$. By analogy, the probability of finding the particle described by Eq. (2.35) at a given point x is proportional to $\psi^2(x)$, as mentioned previously and corresponding to Born's interpretation of wavefunctions (Chapter 1).

The Schrödinger equation for a particle moving in three-dimensional space with potential energy V, is

$$[-(\hbar^2/2m)\nabla^2 + V]\psi = E\psi \tag{2.42}$$

providing the hamiltonian and the wavefunction (hence the total energy) do not change with time. The energy values and the wavefunctions will then describe *stationary states*. However, in the more general case, the system will change with time and the wavefunction $\Psi(q,t)$ will have both a spatial (q represents any spatial coordinate) and a time dependence. Such a wavefunction will contain the information on the time evolution of the observables, within the limits imposed by the indeterminacy principle and in the context of the statistical determinism. This requires that $\Psi(q,t)$ is related with $\partial\Psi(q,t)/\partial t$. The time-dependent Schrödinger equation which must be satisfied by $\Psi(q,t)$ is

$$\hat{H}\Psi(q,t) = i\hbar\,\partial\Psi(q,t)/\partial t. \tag{2.43}$$

In many cases it is possible to separate the time dependence of $\Psi(q,t)$ from its spatial variation: $\Psi(q,t) = \psi(q)f(t)$. Then, if the hamiltonian is time independent ($\hat{H} = \hat{H}$), Eq. (2.43) can be split into two equations:

$$\hat{H}\psi(q) = E\psi(q)$$
$$i\hbar\,df(t)/dt = Ef(t) \tag{2.44}$$

the former being the time-independent wave equation and the latter representing the time evolution. The second equation has the solution

$$f(t) = A\exp(-iEt/\hbar) \tag{2.45}$$

hence the time-dependent solutions of Eq. (2.43) become

$$\Psi(q,t) = \psi(q)e^{-iEt/\hbar}. \tag{2.46}$$

The exponential in Eq. (2.46) is called the *phase factor* of $\Psi(q,t)$. It has no effect on energy or particle distribution. For example, the probability density $\Psi\Psi^*$ is

$$\Psi\Psi^* = \psi e^{-iEt/\hbar}\psi^* e^{iEt/\hbar} = \psi\psi^*. \tag{2.47}$$

Therefore, phase factors can be ignored in dealing with stationary states. This situation is analogous to the case of standing waves in classical mechanics.

It is noted that the complete Schrödinger equation is a second-order differential equation in the spatial coordinates and a first-order differential equation in the variable time. Therefore, it is not rigorously a wave equation (which would require a second derivative with respect to time). On the other hand, the variable time does not enter the equation as an observable but as a parameter to which well-defined values are attributed. Thus, there are no commutation relations involving a time operator. Nevertheless, it is possible to establish an indeterminacy relation involving energy and time, similar to those previously found for position and momentum. If Δt is the lifetime of a given state of the system, there will be an indeterminacy in the energy of such a state:

$$\Delta E \, \Delta t \geq h/4\pi. \tag{2.48}$$

For a stationary state, $\Delta t = \infty$ and $\Delta E = 0$; the energy is known exactly. If transitions between states are allowed, ΔE can be significant and contributes (natural linewidth) to the linewidth of the spectral lines.

2.6 A simple system: translational motion of a particle

Free particle in linear motion

As shown before, the form of the wave equation for this case is simply

$$-(\hbar^2/2m)\, d^2\psi/dx^2 = E\psi \tag{2.49}$$

where E is the kinetic energy of the particle. The general form of the solutions of this differential equation is

$$\psi = Ae^{ikx} + Be^{-ikx} \tag{2.50}$$

with k given by

$$k = (2mE)^{1/2}/\hbar. \tag{2.51}$$

Problem 2.9   Show by substitution that Eq. (2.50) together with Eq. (2.51) satisfy Eq. (2.49).

There is no limitation to the values of parameter k. Hence, the particle can have any kinetic energy

$$E = k^2\hbar^2/2m. \tag{2.52}$$

The energy of a free particle in linear motion, without any boundaries, is *not* quantized.

For $B=0$, the wavefunction (2.50) describes a particle moving in direction $+x$ with linear momentum

$$p = (2mE)^{1/2} = k\hbar \qquad (2.53)$$

whereas, for $A=0$, the motion is in the opposite direction with the same magnitude of linear momentum. In any case, since $\exp(ikx) = \cos kx + i \sin kx$ and an harmonic wave has the form $\cos(2\pi x/\lambda)$ (or $\sin(2\pi x/\lambda)$) it is concluded that the wavefunction is associated with a wavelength $\lambda = 2\pi/k$. The probability density function is simply $\psi\psi^* = Ae^{-ikx}Ae^{ikx} = A^2$ (or B^2), i.e. it is a constant. The position of the particle cannot be specified, in accordance with the fact that its momentum $p=k\hbar$ is well defined.

The de Broglie relation is again reproduced:

$$p = k\hbar = (2\pi/\lambda)(h/2\pi) = h/\lambda. \qquad (2.54)$$

Problem 2.10 Show that the function (2.50) can be expressed as $\psi = C \sin kx + D \cos kx$.

For an extension of this treatment to two dimensions see, for example, ref. 17. In this case, besides the linear momentum (and the energy) as a constant of motion, that is, a quantity that does not change during the motion of the particle, there is also an angular momentum (with respect to the origin of the reference frame) which is another constant of motion.

Free particle in a one-dimensional 'box'

Energy quantization does occur if the particle is confined to a specified region in space. Let us consider a free particle allowed to move along a straight segment between $x=0$ and $x=a$. We refer to a one-dimensional 'box'. This is equivalent to considering that the potential energy is $V=0$ between $x=0$ and $x=a$ (excluding these points), and suddenly becomes $V=\infty$ at the walls (Fig. 2.1).

The general form of the wavefunctions is still given by Eq. (2.50), which is more conveniently written (see Problem 2.10) as

$$\psi = C \sin kx + D \cos kx \qquad (2.55)$$

and the expression for the energies is again given by Eq. (2.52). However, the wavefunctions must now conform to certain conditions. They must vanish for

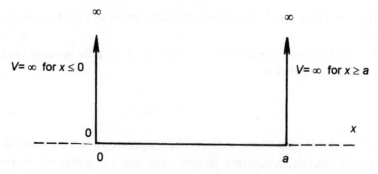

Fig. 2.1 The potential of a free particle in a one-dimensional box.

$x = 0$, because the particle has zero probability of being at that point (*first boundary condition*). Thus,

$$\psi^2(0) = 0$$
$$\psi(0) = 0, \ D = 0$$
(2.56)

and Eq. (2.55) becomes

$$\psi = C \sin kx.$$
(2.57)

In addition (*second boundary condition*) they must vanish for $x = a$:

$$\psi^2(a) = 0$$
$$\psi(a) = 0, \ C \sin ka = 0.$$
(2.58)

Since C cannot be zero, because then ψ would be zero for all points (and the particle would not be in the segment considered), it must be concluded that $\sin ka = 0$ which requires

$$ka = n\pi, \ n = 1, 2, 3, \ldots$$
(2.59)

Using expression (2.52) and replacing k by its value given by Eq. (2.59), we have

$$E = k^2\hbar^2/2m = n^2h^2/8ma^2$$
(2.60)

with

$$n = 1, 2, 3, \ldots$$

called a *quantum number*. As a result of the boundary conditions imposed upon the wavefunctions, reflecting the confinement of the particle, the (kinetic) energies of the particle are quantized.

The value $n = 0$ cannot be considered in Eq. (2.59) (or in Eq. 2.60) because it would imply $k = 0$ and thus $\psi = 0$ and $\psi^2 = 0$ for all points (the particle would not be in the segment considered). This means that, in contrast to classical mechanics, the minimum energy is not zero, but

$$E_1 = h^2/8ma^2 \tag{2.61}$$

which is called the *zero-point energy*.

From the difference between successive energy values

$$\Delta E = E_{n+1} - E_n = (2n+1)h^2/8ma^2 \tag{2.62}$$

it is found that energy separation increases with the quantum number. However, it would always be zero for $a = \infty$, in agreement with the previous conclusion that there is no quantization if there are no boundaries. Expression (2.62) also shows that the energy separations decrease when either a or m increase. Thus, the more macroscopic the system is, the less important is the energy quantization. Classical mechanics is valid for macroscopic bodies.

The forms of the low-energy wavefunctions are shown below:

$$
\begin{aligned}
n = 1 \quad & k = \pi/a \quad & \psi_1 = C \sin[(\pi/a)x] \\
n = 2 \quad & k = 2\pi/a \quad & \psi_2 = C \sin[(2\pi/a)x] \\
n = 3 \quad & k = 3\pi/a \quad & \psi_3 = C \sin[(3\pi/a)x].
\end{aligned}
\tag{2.63}
$$

For $x = a/2$ (midpoint in the segment), ψ_1 has a maximum, ψ_2 has a *node* ($\psi_2 = 0$), ψ_3 has a minimum (sin $3\pi/2 = -1$); the latter has nodes for $x = a/3$ and $x = 2a/3$. Figure 2.2 shows the graphical representations of these functions and the corresponding energy levels.

Problem 2.11 Confirm in Fig. 2.2 that the wavefunctions are orthogonal.

The numerical parameter C in Eq. (2.63) must be chosen so that the wavefunctions are normalized:

$$\int_0^a \psi^2 \, dx = 1. \tag{2.64}$$

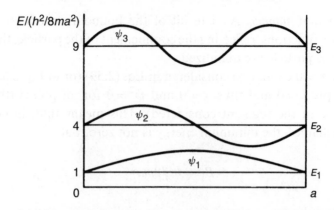

Fig. 2.2 Wavefunctions and energies for a free particle in a one-dimensional box.

That is,

$$C^2 \int_0^a \sin^2 kx \, dx = C^2 a/2 = 1$$

$$C = (2/a)^{1/2}.$$

(2.65)

The form of the wavefunctions is such that they are not destroyed by interference. In fact, the corresponding wavelengths are such that

$$a = n\lambda/2, \quad n = 1, 2, 3, \ldots$$

$$\lambda = 2a/n.$$

(2.66)

Comparison with

$$p = kh/2\pi = nh/2a$$

(2.67)

yields the de Broglie equation $p = h/\lambda$.

It should however be noted that the eigenfunctions of the hamiltonian are not eigenfunctions of the linear momentum operator. Accordingly, a measurement of the momentum does not lead to $p_n^2 = 2mE_n$, but to a probability distribution as shown in Fig. 2.3 (refs. 18 and 19). It can be noted that the most probable momentum is not $\pm p_n$, when the particle is in a state ψ_n except when n becomes large. Then the quantum and classical descriptions are similar.

At this point it is stressed that the Schrödinger equation can also be formulated and solved in momentum space instead of coordinate space.

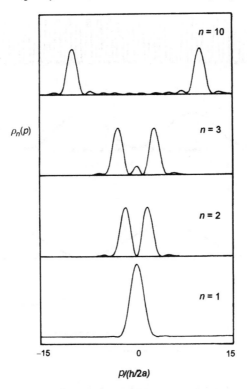

Fig. 2.3 Probability distribution of the linear momentum p for a particle in a one-dimensional box. (Adapted with permission from the *Journal of Chemical Education*, **72**, 148 (1995); copyright 1995, Division of Chemical Education Inc., ref. 18.)

The solutions $\omega(p)$ obtained contain the same information about the physical properties of the system as the eigenfunctions of the hamiltonian $\psi(q)$, the two sets of functions being related through a Fourier transform (ref. 20). We will come back to this point when dealing with electrons, especially in Chapter 3 which deals with one-electron atoms.

Figure 2.2 already shows that the number of nodes increases with n, corresponding to a decrease of λ and an increase of both p and E. The probability density ψ_n^2 varies according to Fig. 2.4. It also depicts some of the positions where the particle can be instantaneously found in many determinations of an ensemble of identical systems equally prepared in such a way that we know them to all be in a given state n. For very high n (high energy), the probability distribution is practically uniform, thus approaching the classical description according to which the averaged residential time of the particle in each position is the same. This is another example of the tendency of quantum-mechanical predictions to approach classical predictions when

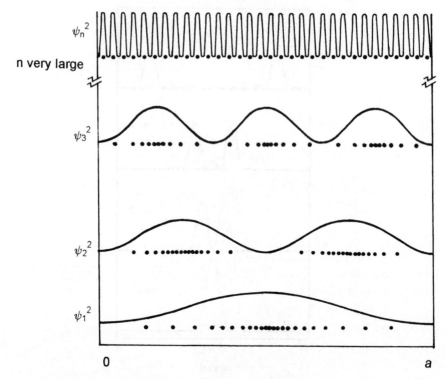

Fig. 2.4 Probability density and particle distribution for a free particle in a one-dimensional box, for different energy levels. (Redrawn from ref. 13.)

one goes from the sub-microscopic to the macroscopic realm: the so-called *correspondence principle*.

The occurrence of a node at the midpoint in the $n=2$ state (and similarly for the states higher in energy) might raise questions about how the particle can get from one side to the other, if the probability of it being found at the midpoint is zero. This question is based on the idea of a trajectory which is contrary to quantum mechanics where what is meaningful is the probability density in each point (and only negative and imaginary values are not acceptable). What we can say is that the probability of the particle being in the left half is 50% and is equal to the probability of being in the right half of the segment. Any classical answer would require successive measurements on the same particle, initially prepared in the $n=2$ state and kept as such during the process, which is impossible because any measurement affects the system.

If we are interested in the probability of finding the particle in an element Δx, then it is never zero, however small it may be, because ψ^2 may be zero at a given point but does not vanish in the neighbouring points in Δx. It has been shown (ref. 21) that around the nodal point $x = a/2$, for $n = 2$, the elementary probability ΔP is

$$\Delta P = (8\pi^2/3a^3)(\Delta x)^3 + \text{higher order terms.} \qquad (2.68)$$

Later in the chapter we will see that once relativistic corrections are introduced there are no longer any nodes.

Brief extensions and tunnelling

Particle in a three-dimensional box

A natural extension of the previous discussion is to consider a particle in the three-dimensional analogue of the square well for the one-dimensional box. Since the motions in the three directions are independent from each other, the hamiltonian in Eq. (2.42) (with $V = 0$) can be written as a sum of operators, one in each variable x, y or z, and ψ can be written as a product of three functions, each one being a function of a different variable: $\psi(x, y, z) = X(x) Y(y) Z(z)$. The problem to solve is then similar to that of the one-dimensional box, taken three times. The energy is the sum of three components similar to Eq. (2.60), involving three quantum numbers:

$$E_{n_x, n_y, n_z} = [(n_x/a_x)^2 + (n_y/a_y)^2 + (n_z/a_z)^2]h^2/8m \qquad (2.69)$$

whereas the wavefunctions are products of three one-variable functions having the same form as Eq. (2.63):

$$\psi(x, y, z) = (8/a_x a_y a_z)^{1/2} \sin[(n_x\pi/a_x)x] \sin[(n_y\pi/a_y)y] \sin[(n_z\pi/a_z)z]. \qquad (2.70)$$

Problem 2.12   Translation is the only molecular motion to consider for a monoatomic gas in order to characterize molar entropies.
(a) *What happens to the translational energy levels when the volume of 1 mol of gas increases?*
(b) *How do energy level separations depend on molecular mass?*
(c) *Relate the answers to (a) and (b) with the changes in molar entropy of a monoatomic gas with volume and molar mass.*

It is clear from Eq. (2.69) that equal values of energy can be obtained using different sets of quantum numbers n_x, n_y and n_z, that is, degenerate eigenfunctions of the hamiltonian exist. Figure 2.5 shows plots of two

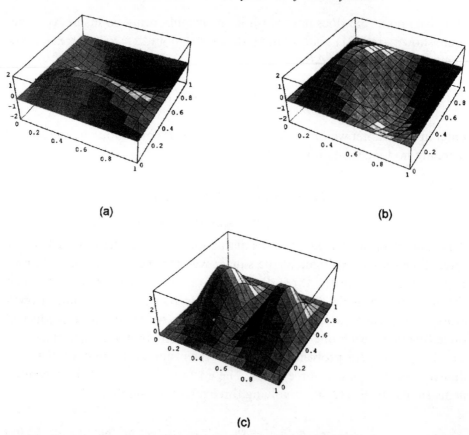

(a)   (b)

(c)

Fig. 2.5 Plots of ψ surfaces corresponding to equal energy ($n_x = 1$, $n_y = 2$, and $n_x = 2$, $n_y = 1$) for a free particle in a square box: (a) and (b). The probability density corresponding to (b) is also shown: (c). (Reprinted with permission from the *Journal of Chemical Education*, **75**, 506 (1998); copyright 1998, Division of Chemical Education Inc., ref. 22.)

degenerate ψ surfaces for a particle in a two-dimensional box (ref. 22) calculated using available software (ref. 23).

Particle on a ring

If, instead of a linear segment, the particle is free to move around a ring of radius R to which it is confined, coordinate x can be replaced by the angle of rotation ϕ, angular velocity $\omega = v/R$ becomes the analogue of linear velocity v, and linear momentum mv gives rise to angular momentum $mvR = mR^2\omega = I\omega$ (I is the moment of inertia). The wave equation is similar to Eq. (2.49), with ϕ instead of x and I instead of m:

$$-(\hbar^2/2I)\,d^2\psi/d\phi^2 = E\psi. \tag{2.71}$$

The solutions have a similar form to the one-dimensional case and

$$E = k^2 \hbar^2 / 2I. \tag{2.72}$$

The boundary conditions have changed: now it is required that ψ (and $d\psi/d\phi$) repeats itself every time ϕ changes by 2π radians. As a consequence, the allowed values of k are $0, 1, 2, 3, \ldots$. In this case, there is no zero-point energy.

Free particle on a spherical surface

There are now two independent angular coordinates, θ and ϕ (besides the fixed distance R to the centre of the sphere) and wavefunctions that can be expressed as products of $\Theta(\theta)$ and $\Phi(\phi)$. The wave equation can be split into two, one for each variable θ and ϕ. Each one of the functions $\Theta(\theta)$ and $\Phi(\phi)$ is subject to boundary conditions, in a similar way to the motion on a circular ring. Accordingly, two quantum numbers arise. The complete solutions $\Theta(\theta)\Phi(\phi)$ are known as *spherical harmonic functions* and the allowed energies are given by an expression that resembles Eq. (2.72):

$$E = \ell(\ell + 1)\hbar^2 / 2mR^2 \qquad \ell = 0, 1, 2, \ldots \tag{2.73}$$

The values $\ell(\ell + 1)\hbar^2$ are found to be the eigenvalues of the operator for the square of the orbital angular momentum $\hat{l}^2$. Thus, ℓ is an orbital angular momentum quantum number. The z component of the angular momentum is given by $m_\ell \hbar$, the quantum number m_ℓ varying between $-\ell$ and $+\ell$, in unit jumps. For plots of the wavefunctions, see, for example, ref. 22.

Energy levels for a particle in a parabolic potential: the harmonic oscillator

The particle is now considered in a one-dimensional potential well given by $V = (1/2)kx^2$ (which is a parabolic function), like a pendulum. There is again only one quantum number, n, and the energies obtained by solving the wave equation are given by

$$E = (n + 1/2)h\nu \qquad n = 0, 1, 2, \ldots \tag{2.74}$$

where

$$\nu = (1/2\pi)\sqrt{k/m} \tag{2.75}$$

is the classical expression for the frequency of oscillation.

It is noted that, as opposed to the case of the free particle in a box of fixed width, the energies vary with n rather than n^2. This can be rationalized by thinking of the particle as occupying successively bigger boxes as its energy increases – in agreement with the fact that a pendulum swings over a longer trajectory if it has a higher energy – and recalling that, as a one-dimensional box is made wider, the separation between energy levels decreases. This counteracts the n^2 dependence previously found.

Free particle in a one-dimensional box with one finite wall: tunnelling

Returning to the one-dimensional box of constant width, if the potential does not increase suddenly to infinity at one of the walls then the wavefunction does not vanish there. Due to the continuity requirement of the wavefunction, it decreases exponentially to zero inside the wall of finite height. Therefore, there is a non-zero probability that the particle will penetrate the wall, although its kinetic energy is lower than the potential barrier (Fig. 2.6). This effect is called *quantum-mechanical tunnelling*.

Figure 2.7 shows the situation encountered if one of the barriers has both a finite height and a finite width. Now, the wavefunction does not vanish inside the barrier and the particle will be described in the outer region, where $V = 0$, as it is inside the box.

It is clear that this effect has no classical counterpart. Classically, a particle with kinetic energy lower than a given potential barrier cannot go through; it would have negative kinetic energy inside the barrier. In fact, for a given value E – which for $V = 0$ inside the box and $V = \infty$ at the walls

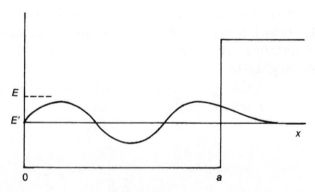

Fig. 2.6 Tunnelling effect for a free particle in a one-dimensional box with wall of finite height. The broken line represents the corresponding energy level should both potential barriers be infinite (see, for example, ref. 13).

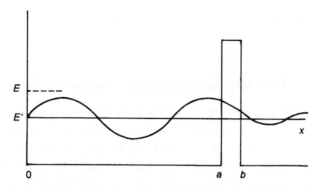

Fig. 2.7 Tunnelling effect for a free particle in a one-dimensional box with one wall of finite height and finite width. The broken line represents the corresponding energy level should both potential barriers be infinite (see also ref. 13).

would be only kinetic energy – inside the barrier we have a contribution V ($V > 0$), with $E < V$ (total energy smaller than potential energy). This result is acceptable in quantum mechanics, because only the average value of an observable has physical meaning, and this average takes the 'entire' wavefunction into account. The observable kinetic energy value is never negative.

Problem 2.13   Consider that, for the region inside the barrier in Fig. 2.7 ($V > 0$, $E < V$), the wave equation (2.49) will have $(E - V) < 0$ instead of E.
(a) *Show that the wavefunctions inside the barrier involve positive exponents*

$$\psi = Ae^{k'x} + Be^{-k'x}$$

with

$$k' = [2m(V - E)]^{1/2}/\hbar.$$

(b) *Conclude that $A = 0$, because, otherwise, ψ would go to infinity as x increases.*

It can be expected that the higher and the thicker the barrier in Fig. 2.7 the less is the probability that the particle be found outside the box. Equally expected is that the heavier the particle the smaller that probability. In fact, the exponential variation of ψ inside the barrier (see Problem 2.13)

$$\psi = Be^{-k'x} \tag{2.76}$$

depends on k' which, as in Eq. (2.51), increases with m. This explains why electrons can tunnel much more efficiently than protons. The differences in the rates of some reactions when protium is replaced by deuterium is due to the different capacity of tunnelling. A recent practical application of tunnelling electrons is scanning tunnelling microscopy.

2.7 Relativity theory, quantum mechanics and spin

The time-dependent Schrödinger equation (2.43) presents a serious problem from the point of view of relativity theory: it does not treat space and time in a symmetric way, because second-order derivatives of the wavefunction with respect to spatial coordinates are accompanied by a first-order derivative with respect to time. One way out, as actually proposed by Schrödinger and later known as the Klein–Gordon equation, would be to have also second-order derivatives with respect to time. However, that would lead to a total probability for the particle under consideration which would be a function of time, and to a variation of the number of particles of the universe (which, at the time, was completely unacceptable). In 1928, Dirac sought the solution for this problem, by accepting first-order derivation in the case of time and forcing the spatial derivatives to also be first order. This requires the wavefunction to have four components (functions of the spatial coordinates alone), often called a four-component 'spinor'.

Two of the components of ψ in Dirac's theory correspond to negative values for the kinetic energy. To overcome this difficulty, Dirac proposed that all states of negative kinetic energy in the universe would be filled and that we only have experience of those particles having positive kinetic energies. This idea would obtain experimental support with the discovery of the positron (or antiparticle of the electron) corresponding to the hole (of positive charge) left by an electron that had been excited from those states to levels of positive kinetic energy.

The occurrence of two degenerate solutions (of positive kinetic energy) for the electron, which are split in the presence of a magnetic field, can be interpreted by attributing an intrinsic magnetic moment to the electron, with two possible and opposed orientations in a magnetic field. This is in accord with the suggestion made in 1925 by the Dutch-American physicists G. Uhlenbeck and S. Goudsmit and developed into a consistent theory by the Austrian physicist, later naturalized American, Wolfgang Pauli (1900–1958, 1945 Nobel laureate in Physics), to account for certain features of atomic spectra, that the electron has a *spin* angular momentum characterized by a quantum number 1/2 (magnetic quantum numbers ±1/2).

In addition, Dirac's theory provides a direct explanation for the fact that the electron magnetic dipole moment is about twice the value expected classically on the basis of a spherical charged particle rotating around one of its axes. Such an explanation does not require any rotation of the electron. The only motion considered in the relativistic quantum mechanics of Dirac is translational motion. Should an intrinsic rotation also be included, the agreement with experiment would be lost.

It has been noted that spin occurs naturally in Dirac's theory not because it is a relativistic phenomenon, in the usual sense of this expression, but because the wavefunction has the correct properties. In some interpretations, it is not even a property of the electron (and other particles) but a manifestation of the interaction between the electron and the vacuum, whose field acquires angular momentum, besides the angular momentum the particle may have (see ref. 24).

For the case of an electron, or any other particle of spin 1/2 (a *fermion*) considered free in a one-dimensional (the z direction, say) potential well of infinite walls, the time-independent relativistic wave equation

$$\hat{H}\psi = E\psi$$

takes into account that now the kinetic energy operator is

$$\alpha_z \hat{p}_z c + \beta mc^2 \qquad \hat{p}_z = -i\hbar \, \partial/\partial z \qquad (2.77)$$

where c is the speed of light and α_z and β are 2×2 matrices:

$$\alpha_z = \begin{bmatrix} 0 & \sigma_z \\ \sigma_z & 0 \end{bmatrix} \qquad \beta = \begin{bmatrix} I & 0 \\ 0 & I \end{bmatrix}. \qquad (2.78)$$

In turn, I is the 2×2 unity matrix

$$I = \begin{bmatrix} 1 & 0 \\ 0 & 1 \end{bmatrix} \qquad (2.79)$$

and σ_z is another 2×2 matrix (the Pauli matrix):

$$\sigma_z = \begin{bmatrix} 1 & 0 \\ 0 & -1 \end{bmatrix}. \qquad (2.80)$$

Problem 2.14   *Identify the Pauli matrix above in the following form of the spin angular momentum along the z direction*

$$[S_z] = (\hbar/2) \begin{bmatrix} 1 & 0 \\ 0 & -1 \end{bmatrix}$$

considering the two allowed S_z values for $m_s = 1/2$ and $m_s = -1/2$.

One of the consequences of having a first-order derivative with respect to z and not $d^2\psi/dz^2$ is that the wavefunctions do not vanish at the walls of the box. In fact, the probability density is never zero for any point of the

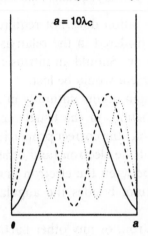

Fig. 2.8 Probability densities for a relativistic particle in an one-dimensional box (three lowest-energy levels) (see also ref. 25).

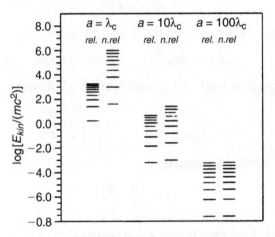

Fig. 2.9 Energy diagrams for a free particle in an one-dimensional potential well, obtained with and without the relativistic correction (see also ref. 25).

box; if one component of ψ vanishes at a given point, another component is non-zero there.

Figure 2.8 shows the total probability density for the lowest-energy functions calculated for the case $a = 10\lambda_c$ where $\lambda_c = \hbar/mc$ is the Compton wavelength for the particle, divided by 2π (for the electron, it is $\lambda_c = 0.4\,\text{pm}$).

Figure 2.9 compares the energy level diagrams in the non-relativistic approximation and in Dirac's theory. The latter values are always smaller than the non-relativistic counterparts. For larger dimensions of the box, $a = 100\lambda_c$, the relativistic corrections are negligible.

3

One-electron atoms: atomic orbitals

3.1 Wave equation and angular momentum

In the Schrödinger equation

$$\hat{H}\psi = E\psi \tag{3.1}$$

for the hydrogen atom, or any other monoelectronic atom (ion) of nuclear charge Ze, the hamiltonian has the following form:

$$\hat{H} = -(\hbar^2/2m_e)\nabla_e^2 - (\hbar^2/2m_N)\nabla_N^2 + V. \tag{3.2}$$

In Eq. (3.2), m_e and m_N represent the electron mass and the mass of the nucleus, respectively; V is the potential energy corresponding to the electron–nucleus attraction:

$$V = -\pounds Ze^2/r \tag{3.3}$$

($\pounds = 1/4\pi\varepsilon_0 = 8.99 \times 10^9 \,\mathrm{J\,m\,C^{-2}}$ is the electromagnetic constant). The laplacians ∇_e^2 and ∇_N^2 are the known operators (see expression (2.27)) involving, respectively, the electron and the nucleus coordinates:

$$\nabla_e^2 = \partial^2/\partial x_e^2 + \partial^2/\partial y_e^2 + \partial^2/\partial z_e^2$$
$$\nabla_N^2 = \partial^2/\partial x_N^2 + \partial^2/\partial y_N^2 + \partial^2/\partial z_N^2. \tag{3.4}$$

Considering the centre of mass of the atom, O, and the reduced mass of the system defined as $\mu_r = m_e\, m_N/(m_e + m_N)$, Eq. (3.2) can be written:

$$\hat{H} = -(\hbar^2/2M)\nabla_O^2 - (\hbar^2/2\mu_r)\nabla^2 + V \tag{3.5}$$

45

(see, for example, ref. 12). Taking $M = m_e + m_N$ and using second derivatives referred to the coordinates of the centre of mass O, the first term corresponds to the kinetic energy of the atom as a whole (a free particle). The second term refers to the kinetic energy of the electron inside the atom. Rigorously, this term represents the motion of a particle of mass μ_r around the centre of mass O. Since it is this motion that is of most importance, the wave equation to solve is

$$-(\hbar^2/2\mu_r)\nabla^2\psi + V\psi = E\psi \tag{3.6}$$

which can be converted into

$$-(\hbar^2/2m)\nabla^2\psi + V\psi = E\psi \tag{3.7}$$

or

$$\nabla^2\psi + (2m/\hbar^2)(E - V)\psi = 0 \tag{3.8}$$

by making the reasonable approximation $m_e + m_N = m_N$ and, consequently, $\mu_r = m_e = m$ (electron mass from now onwards represented by m). With this approximation, we can refer to the motion of the electron about the nucleus.

By making use of the relations between cartesian coordinates and spherical polar coordinates (Fig. 3.1),

$$x = r\sin\theta\cos\phi$$
$$y = r\sin\theta\sin\phi$$
$$z = r\cos\theta$$
$$r^2 = x^2 + y^2 + z^2$$
$$\text{for} \quad -\infty < x, y, z < +\infty$$
$$0 < r < \infty$$
$$0 < \theta < \pi$$
$$0 < \phi < 2\pi,$$

a change of variables can be effected:

$$\partial/\partial x = \partial(\partial r/\partial x)/\partial r$$
$$\partial^2/\partial x^2 = (\partial r/\partial x)\partial^2/\partial r\partial x + (\partial^2 r/\partial x^2)\partial/\partial r, \text{ etc.}$$

with

$$\partial r/\partial x = \partial(x^2 + y^2 + z^2)^{1/2}/\partial x = x/r = \sin\theta\cos\phi. \tag{3.9}$$

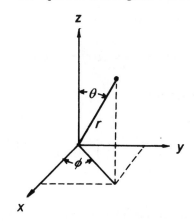

Fig. 3.1 Spherical polar coordinates.

Equation (3.8) then converts into

$$\frac{1}{r^2}\frac{\partial}{\partial r}\left(r^2\frac{\partial\psi}{\partial r}\right)+\frac{1}{r^2\sin\theta}\frac{\partial}{\partial\theta}\left(\sin\theta\frac{\partial\psi}{\partial\theta}\right)+\frac{1}{r^2\sin\theta}\frac{\partial^2\psi}{\partial\phi^2}+\frac{2m}{\hbar^2}(E-V)\psi=0.$$

$$(3.10)$$

Problem 3.1   Consider a free particle ($V=0$) moving in a plane xy.
(a) Write down the Schrödinger equation.
(b) Relate the coordinates x and y to r and ϕ of Fig. 3.1.
(c) Show that the wave equation can be written as

$$\hbar^2\,d^2\psi/d\phi^2=-2IE\psi$$

if r is constant (particle rotating in the xy plane) and with $I=mr^2$ (moment of inertia) (see page 38).
(d) Knowing that the boundary condition $\psi(\phi)=\psi(2\pi+\phi)$ leads to energy quantization

$$E=m_\ell^2\hbar^2/2I\qquad m_\ell=0,\pm1,\pm2,\dots$$

and that $E=J^2/2I$, show that the angular momentum (only in z direction) is $J_z=m_\ell\hbar$.

We can then explore the possibility of expressing the wavefunction by a product of two functions, one only dependent on *r* and the other a function of the angular coordinates θ and ϕ,

$$\psi(r, \theta, \phi) = R(r) \times Y(\theta, \phi). \tag{3.11}$$

As a result Eq. (3.10) leads to:

$$\frac{1}{R}\frac{d}{dr}\left(r^2\frac{dR}{dr}\right) + \frac{2m}{\hbar^2}(E - V)r^2 = -\frac{1}{Y\sin\theta}\frac{\partial}{\partial\theta}\left(\sin\theta\frac{\partial Y}{\partial\theta}\right) - \frac{1}{Y\sin^2\theta}\frac{\partial^2 Y}{\partial\phi^2}. \tag{3.12}$$

Since the left side of this equation is a function only of r (i.e. it does not depend on the values given to θ and ϕ) and the right side is a function only of θ and ϕ (i.e. it does not depend on the value of r), this equality requires that both members of the equation are equal to a constant, μ. Rearranging leads to:

$$\frac{1}{\sin\theta}\frac{\partial}{\partial\theta}\left(\sin\theta\frac{\partial Y}{\partial\theta}\right) + \frac{1}{\sin^2\theta}\frac{\partial^2 Y}{\partial\phi^2} + \mu Y = 0 \tag{3.13}$$

$$\frac{1}{r^2}\frac{d}{dr}\left(r^2\frac{dR}{dr}\right) + \left[\frac{2m}{\hbar^2}(E - V) - \frac{\mu}{r^2}\right]R = 0. \tag{3.14}$$

Equation (3.13) has the same form as

$$(-1/\hbar^2)\hat{l}^2 Y(\theta, \phi) = -\mu Y(\theta, \phi) \tag{3.15}$$

where $\hat{l}$ is the orbital angular momentum operator in three dimensions (see page 26). The eigenvalue of $\hat{l}^2$

$$\mu\hbar^2 = \ell(\ell + 1)\hbar^2 \tag{3.16}$$

satisfies

$$\hat{l}^2 Y(\theta, \phi) = \ell(\ell + 1)\hbar^2 Y(\theta, \phi) \tag{3.17}$$

with ℓ being an integer and positive number (or zero). This is called the *orbital angular momentum quantum number*. This quantum number is related to the eigenvalue of $\hat{l}_z$

$$\hat{l}_z Y(\theta, \phi) = m_\ell\hbar\, Y(\theta, \phi) \tag{3.18}$$

through

$$\ell \geq m_\ell$$

where m_ℓ is an integer number. This means:

$$m_\ell = -\ell, -\ell + 1, \ldots, 0, \ldots, \ell - 1, \ell. \tag{3.19}$$

The quantum number m_ℓ is called the *magnetic quantum number*, because the z direction is only unequivocally defined in terms of the direction of a magnetic field applied to the atom.

Problem 3.2 *Consider a free particle ($V=0$) moving on a spherical surface.*
(a) *Write out the expression for its angular momentum (see Problem 3.1).*
(b) *Using the classical equation $E = J^2/2I$, where $I = mr^2$ is the moment of inertia, show that the energy is quantized according to:*

$$E = \ell(\ell + 1)\hbar^2/2I, \quad \ell = 0, 1, 2, \ldots$$

(*see also page* 39).

Problem 3.3 *Confirm that the following $Y(\theta, \phi)$ functions (spherical harmonics) satisfy Eq. (3.13):*

ℓ	m_ℓ	$Y(\theta, \phi)$
0	0	$(1/4\pi)^{1/2}$
1	0	$(3/4\pi)^{1/2} \cos\theta$
1	± 1	$\mp(3/8\pi)^{1/2} \sin\theta\, e^{\pm i\phi}$

Equation (3.14) becomes

$$\frac{d^2 R}{dr^2} + \frac{2}{r}\frac{dR}{dr} + \left[\frac{2m}{\hbar^2}(E - V) - \frac{l(l+1)}{r^2}\right]R = 0 \tag{3.20}$$

which can be solved by introducing a parameter n, such that

$$E = -(Z^2 me^4/8\varepsilon_0^2 h^2)/n^2 \tag{3.21}$$

and recognizing that only well-behaved $R(r)$ solutions are acceptable (see, for example, ref. 26). (By 'well-behaved' we mean functions that are finite, have no discontinuities (or violent changes of curvature) and which go to zero at infinity.) This follows because the electron density, which is given by the square of the wavefunction (or the product of the wavefunction by its complex conjugate), cannot be infinite at any point and must change smoothly and continuously with r; also, it must vanish for $r=\infty$. With these conditions, n must be an integer and

$$n \geq \ell + 1$$

hence

$$n = 1, 2, 3, \ldots, \quad \ell < n - 1. \tag{3.22}$$

The energy quantization for the electron of any monoelectronic atom, which corresponds to the discrete values of parameter n in expression (3.21), is thus a direct consequence of a set of reasonable mathematical conditions imposed upon the radial part, $R(r)$, of the wavefunction ψ. The parameter n is called the *principal quantum number*.

Problem 3.4   Confirm that Eq. (3.21) reproduces the expression (1.8) for the electron energy in the H *atom first established by Bohr, by making* $k = 1/4\pi\varepsilon_0 = 1$.

If the angular momentum of the electron is represented by a vector $\mathbf{l}$ of magnitude $[\ell(\ell+1)]^{1/2}h/2\pi$, the z component – along a pre-established direction which can only be unequivocally defined in the presence of an applied field – has quantized magnitude $m_\ell h/2\pi$. This requires a specific orientation of the angular momentum vector with respect to the z direction. For example, for $\ell = 1$, we have

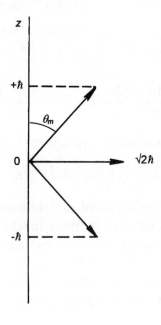

or, rather,

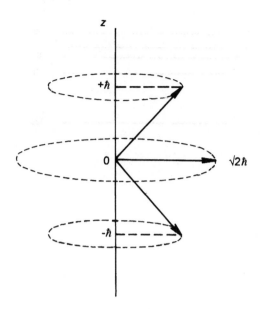

because the actual orientation of the angular momentum vector on a conical surface of angle θ_m cannot be specified. We recall from page 27 that the x and y components of $\mathbf{l}$ are not observable simultaneously with $\mathbf{l}$ and l_z.

3.2 Atomic orbitals

Each energy value allowed for the electron of a monoelectronic atom corresponds to an integral value of the principal quantum number n. In addition, for each value of n there are n values of the orbital angular momentum quantum number ℓ:

$$
\begin{aligned}
n &= 1 & \ell &= 0 \\
n &= 2 & \ell &= 0, 1 \\
n &= 3 & \ell &= 0, 1, 2 \\
&\ \vdots & &\ \vdots \\
n & & \ell &= 0, 1, 2, \ldots, n-1
\end{aligned}
$$

and for each ℓ value there are $2\ell + 1$ values for the magnetic quantum number m_ℓ:

$$
\begin{aligned}
\ell &= 0 & m_\ell &= 0 \\
\ell &= 1 & m_\ell &= -1, 0, +1 \\
\ell &= 2 & m_\ell &= -2, -1, 0, +1, +2 \\
&\ \vdots & &\ \vdots
\end{aligned}
$$

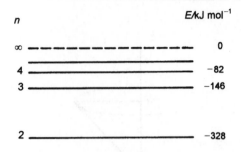

Fig. 3.2 Energy values of the electron in the H atom (reference state: $n=\infty$, $E=0$).

Therefore, for each energy of the electron, there will be

$$\sum_{0}^{n-1}(2\ell + 1) = n^2$$

eigenfunctions characterized by a specific set of quantum numbers n, ℓ and m_{ℓ}:

$$\psi_{n,\ell,m_{\ell}} = R_{n,\ell}Y_{\ell,m_{\ell}}. \tag{3.23}$$

These functions are called the *atomic orbitals*.

Figure 3.2 shows the energy diagram for the H atom.

For the lowest energy level ($-1312\,\text{kJ}\,\text{mol}^{-1}$) there is only one wave-function, $\psi_{1,0,0}$. This is called a 1s orbital. The electron is described by this wavefunction; it is said to 'occupy' a 1s orbital. For $n=2$ (energy $-328\,\text{kJ mol}^{-1}$), there are four atomic orbitals:

$$\psi_{2,0,0} \qquad \text{2s orbital}$$

$$\left.\begin{array}{l} \psi_{2,1,+1} \\ \psi_{2,1,0} \\ \psi_{2,1,-1} \end{array}\right\} \quad \text{2p orbitals.} \tag{3.24}$$

Problem 3.5 *Show that, for $n=3$, there are nine orbitals (including five 3d orbitals) and that, for $n=4$, the total number of orbitals is 16 (including seven 4f orbitals).*

Table 3.1 *Orbitals for monoelectronic atoms, up to $n=3$*

$n=1,\ \ell=0,\ m_\ell=0$	$\psi_{1s} = \dfrac{1}{\sqrt{\pi}}\left(\dfrac{Z}{a_0}\right)^{3/2} e^{-\rho}$	$\left(\rho = \dfrac{Z}{a_0}r\right)$
$n=2,\ \ell=0,\ m_\ell=0$	$\psi_{2s} = \dfrac{1}{4\sqrt{2\pi}}\left(\dfrac{Z}{a_0}\right)^{3/2}(2-\rho)e^{-\rho/2}$	
$n=2,\ \ell=1,\ m_\ell=0$	$\psi_{2p_z} = \dfrac{1}{4\sqrt{2\pi}}\left(\dfrac{Z}{a_0}\right)^{3/2}\rho\, e^{-\rho/2}\cos\theta$	
$n=2,\ \ell=1,\ m_\ell=\pm 1$	$\psi_{2p_x} = \dfrac{1}{4\sqrt{2\pi}}\left(\dfrac{Z}{a_0}\right)^{3/2}\rho\, e^{-\rho/2}\sin\theta\cos\phi$	
	$\psi_{2p_y} = \dfrac{1}{4\sqrt{2\pi}}\left(\dfrac{Z}{a_0}\right)^{3/2}\rho\, e^{-\rho/2}\sin\theta\sin\phi$	
$n=3,\ \ell=0,\ m_\ell=0$	$\psi_{3s} = \dfrac{2}{81\sqrt{3\pi}}\left(\dfrac{Z}{a_0}\right)^{3/2}(27-18\rho+2\rho^2)e^{-\rho/3}$	
$n=3,\ \ell=1,\ m_\ell=0$	$\psi_{3p_z} = \dfrac{2}{81\sqrt{\pi}}\left(\dfrac{Z}{a_0}\right)^{3/2}(6\rho-\rho^2)e^{-\rho/3}\cos\theta$	
$n=3,\ \ell=1,\ m_\ell=\pm 1$	$\psi_{3p_x} = \dfrac{2}{81\sqrt{\pi}}\left(\dfrac{Z}{a_0}\right)^{3/2}(6\rho-\rho^2)e^{-\rho/3}\sin\theta\cos\phi$	
	$\psi_{3p_y} = \dfrac{2}{81\sqrt{\pi}}\left(\dfrac{Z}{a_0}\right)^{3/2}(6\rho-\rho^2)e^{-\rho/3}\sin\theta\sin\phi$	
$n=3,\ \ell=2,\ m_\ell=0$	$\psi_{3d_{z^2}} = \dfrac{1}{81\sqrt{6\pi}}\left(\dfrac{Z}{a_0}\right)^{3/2}\rho^2 e^{-\rho/3}(3\cos^2\theta-1)$	
$n=3,\ \ell=2,\ m_\ell=\pm 1$	$\psi_{3d_{xz}} = \dfrac{\sqrt{2}}{81\sqrt{\pi}}\left(\dfrac{Z}{a_0}\right)^{3/2}\rho^2 e^{-\rho/3}\sin\theta\cos\theta\cos\phi$	
	$\psi_{3d_{yz}} = \dfrac{\sqrt{2}}{81\sqrt{\pi}}\left(\dfrac{Z}{a_0}\right)^{3/2}\rho^2 e^{-\rho/3}\sin\theta\cos\theta\sin\phi$	
$n=3,\ \ell=2,\ m_\ell=\pm 2$	$\psi_{3d_{x^2-y^2}} = \dfrac{1}{81\sqrt{2\pi}}\left(\dfrac{Z}{a_0}\right)^{3/2}\rho^2 e^{-\rho/3}\sin^2\theta\cos 2\phi$	
	$\psi_{3d_{xy}} = \dfrac{1}{81\sqrt{2\pi}}\left(\dfrac{Z}{a_0}\right)^{3/2}\rho^2 e^{-\rho/3}\sin^2\theta\sin 2\phi$	

Table 3.1 lists the mathematical form of the atomic orbitals up to $n = 3$ (where $a_0 = 4\pi\varepsilon_0\hbar^2/me^2 = 52.9\,\text{pm}$ is the Bohr radius, equal to the radius of the first orbit in Bohr's model). Real functions are indicated. We will see later that for example the 2p orbitals corresponding, respectively, to $m_\ell = +1$ and $m_\ell = -1$ are complex functions which can be replaced by real functions for many (but not all) purposes.

Problem 3.6   Distinguish the radial, R(r), and the angular parts, Y(θ, ϕ) of the various p *orbitals in Table 3.1 and compare the latter with the spherical harmonics mentioned in Problem 3.3.*

s *Orbitals*

It is noted that all s orbitals are functions only of r (and not of θ and ϕ), i.e. each s function has the same value for all points of space at a given distance r from the nucleus. They have *spherical symmetry*. We see that $\psi \rightarrow 0$ as $r \rightarrow \infty$, through a factor $e^{-\rho/n}$, where $\rho = Zr/a_0$. For 1s, this is the only r-dependent factor, and the function decreases exponentially on increasing the electron–nucleus distance (Fig. 3.3).

The probability density function ψ_{1s}^2 can be represented as shown in Fig. 3.4(a) for two dimensions. An alternative graphical representation is given in Fig. 3.4(c), where the more dense cloud close to the centre simulates a region of higher electron probability density. Figure 3.4(b) depicts, via circular contours, some of the infinite number of concentric spherical isodensity surfaces. Schematically, the 1s orbital is often represented by a circle around the nucleus, with a radius large enough so that the probability of the electron being found outside the corresponding spherical surface is comparatively small (less than 10%).

For a 2s orbital, the factor $(2 - \rho)$ vanishes for $\rho = 2$ (that is, $r = 2a_0$), and the spherical surface of radius $r = 2a_0$ is a nodal surface (Fig. 3.5). For a 3s orbital, there are two nodal surfaces. Those figures, picturing the electron probability density of s orbitals, illustrate that the maximum lies at the position of the nucleus. It can be seen from the corresponding expressions in Table 3.1 that this maximum varies with Z^3.

Problem 3.7   Determine the radii for the nodal surfaces of the 3s orbital of the hydrogen atom.

An interesting quantity we need to consider is the average distance between the electron and the nucleus for each orbital. This is defined as the

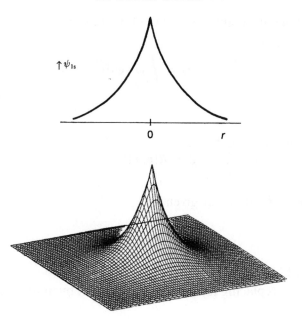

Fig. 3.3 Two- and three-dimensional representations of orbital ψ_{1s} (refs. 27 and 28).

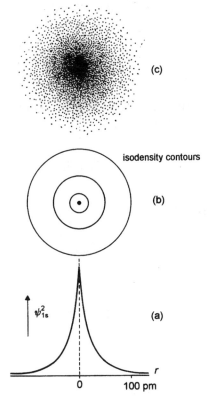

Fig. 3.4 Different graphical ways of representing ψ_{1s}^2.

expectation value of r computed for the function ψ:

$$\langle r \rangle = \int \psi^* r \psi \, dv \tag{3.25}$$

with

$$\psi = R(r) Y(\theta, \phi) \tag{3.26}$$

and the element of volume dv given by

$$dv = r^2 \sin \theta \, dr \, d\theta \, d\phi \tag{3.27}$$

as is shown in Fig. 3.6.

For s orbitals, replacing the angular component by unity, expression (3.25) becomes

$$\langle r \rangle = \int RrR \, dv = \int R^2 r^3 \sin \theta \, dr \, d\theta \, d\phi = \int_0^{2\pi} d\phi \times \int_0^{\pi} \sin \theta \, d\theta \times \int_0^{\infty} R^2 r^3 \, dr$$

$$= 2\pi \times 2 \times \int_0^{\infty} R^2 r^3 \, dr. \tag{3.28}$$

For the 1s orbital of H, using $R(r)$ of Table 3.1 (and considering $\int_0^{\infty} x^n e^{-ax} \, dx = n!/a^{n+1}$) yields

$$\langle r \rangle_{1s} = 3a_0/2, \tag{3.29}$$

while for the 2s orbital

$$\langle r \rangle_{2s} = 6a_0. \tag{3.30}$$

These values are related to each other by a factor of 4, as are the corresponding energies, and are greater by a factor of 3/2 than the radii of the corresponding orbit in the Bohr model.

The general expression for any orbital of a hydrogen-like atom is

$$\langle r \rangle = (n^2 a_0/Z)(3/2 - \ell(\ell+1)/2n^2). \tag{3.31}$$

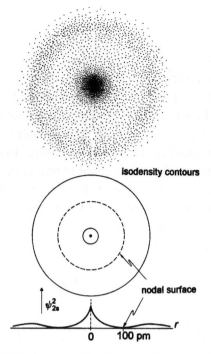

Fig. 3.5 Different graphical ways of representing ψ_{2s}^2.

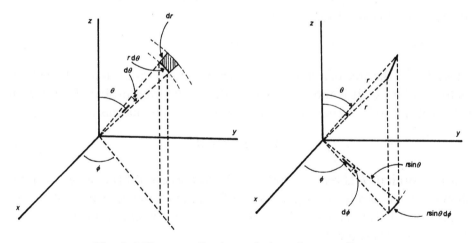

Fig. 3.6 Element of volume dv in polar coordinates.

Problem 3.8   Compare the values for $\langle r \rangle_{2s}$ and $\langle r \rangle_{2p}$ for the hydrogen atom.

It can be shown that

$$\langle r \rangle_{2s} \neq \langle r \rangle_{2p} \tag{3.32}$$

which seems to be at variance with

$$E_{2s} = E_{2p}. \tag{3.33}$$

However, it should be remembered that the average potential energy depends on the average value of $1/r$ and not on the average value of r (or its reciprocal). The same applies to the kinetic energy, hence to the total energy, through the theorem known as the *virial theorem*: for a system with inverse square forces, the average potential energy is equal to minus twice the average kinetic energy, hence the total energy is both equal to minus the kinetic energy and to half the average potential energy (for H, the total electronic energy, the average kinetic energy, and the average potential energy have the forms $-1/(2n^2)$, $1/(2n^2)$, and $-1/n^2$, respectively). Thus, to justify the degeneracy of 2s and 2p orbitals, the following equality is fundamental:

$$\langle 1/r \rangle_{2s} = \langle 1/r \rangle_{2p}. \tag{3.34}$$

A further important quantity to be defined is the r value, independent of the angular positions, for which the electron probability is a maximum, that is, the spherical surface where the electron is most likely to be found. We are now considering a *radial distribution of the probability*, or 'radial probability', which is given by

$$P = 4\pi r^2 \psi^2 \tag{3.35}$$

where $4\pi r^2$ is the area of a spherical surface of radius r (and ψ is a real function, as for s orbitals). It is more realistic to define the total probability for each spherical shell of thickness dr at a distance r from the nucleus as

$$P \, dr = 4\pi r^2 \psi^2 \, dr. \tag{3.36}$$

For a 1s orbital, P is zero for both $r=0$ and $r=\infty$ (in this case, $\psi^2 = 0$) but shows a maximum for $r = a_0$, as can be found from the derivative of P with respect to r. This is shown in Fig. 3.7 (the radial distribution function for the 2s orbital is also shown).

For the 1s orbital of any monoelectronic atom, we have

$$dP/dr = (4Z^3/a_0)(2r - 2Zr^2/a_0)\exp(-2Zr/a_0)$$

from which we get

$$r(P_{max}) = a_0/Z. \tag{3.37}$$

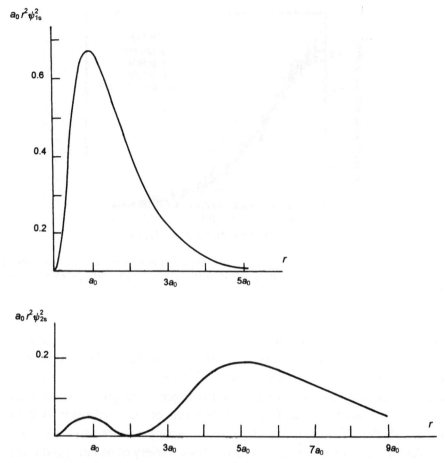

Fig. 3.7 Radial distribution functions $r^2\psi^2$ for the 1s and 2s orbitals of H.

Problem 3.9 *Show that, for a 2s orbital, $r(P_{max}) = (3 + \sqrt{5})a_0/Z$.*

As briefly mentioned in Chapter 2, the Schrödinger equation can be expressed and solved in momentum space yielding $w(p)$ functions instead of atomic orbitals $\psi(r)$. The formal equivalence between these two kinds of functions is expressed by a Fourier transform, which allows one to generate $w(p)$ from $\psi(r)$ and vice versa (refs. 20 and 29):

$$w(p) = (2\pi)^{-3/2} \int \exp(-i\,pr)\psi(r)\,dr$$

$$\psi(r) = (2\pi)^{-3/2} \int w(p)\exp(i\,pr)\,dp.$$

(3.38)

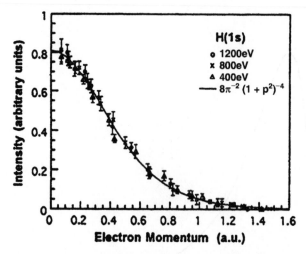

Fig. 3.8 Comparison of theoretical and experimental momentum distribution functions for the hydrogen atom ground state. (Reprinted with permission from ref. 29.)

It has been shown that, under appropriate conditions, the momentum distribution $|\omega(p)|^2$ for an individual electronic state is directly measured by electron-momentum spectroscopy (ref. 29). Figure 3.8 compares the experimental momentum distribution for the hydrogen atom ground state with the function calculated by the Fourier transform of the hydrogen 1s orbital. In general, the electron-momentum spectroscopy results serve to evaluate wavefunctions at various levels of theory for a variety of atomic (and molecular) systems.

p *Orbitals*

An electron in an s orbital has no net angular momentum ($\ell = 0$). However, for p orbitals there is an angular momentum of magnitude $[\ell(\ell+1)]^{1/2}\hbar = \sqrt{2}\,\hbar$. This is classically in agreement with a small amplitude of the p functions close to the nucleus; they vanish for $r = 0$. The $2p_z$ orbital is always zero for $\cos\theta = 0$, that is for any point in the xy plane (*nodal plane*) and has the z axis as an axis of cylindrical symmetry. By defining angles θ_x and θ_y with respect to the x and y axes in the same way as $\theta(\theta_z)$ was defined relative to the z axis, it can be shown that

$$\sin\theta\cos\phi = \cos\theta_x$$
$$\sin\theta\sin\phi = \cos\theta_y. \tag{3.39}$$

Thus, the $2p_x$ and $2p_y$ orbitals are identical to $2p_z$ but instead have the x and y axes as symmetry axes.

Figure 3.9 shows how ψ_{2p_z} varies along the z axis (this is a plot of $R(r)$, because $\theta = 0$). The function ψ_{2p_z} is positive for $z > 0$ and negative for $z < 0$.

Figure 3.10 shows different representations of $\psi^2_{2p_z}$. The isoprobability contour plots for a plane containing the z axis are shown. A rotation around this axis gives rise to the corresponding isoprobability surfaces. The inner curves correspond to higher probability densities.

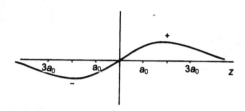

Fig. 3.9 Plot of ψ_{2p_z} for points along the z axis.

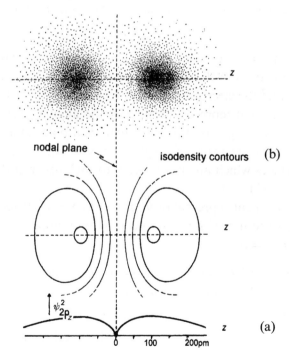

Fig. 3.10 Different graphical ways of representing $\psi^2_{2p_z}$: (a) for points along the z axis; and (b) for a plane containing that axis.

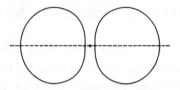

Fig. 3.11 Schematic representation of a 2p orbital by a contour plot which by rotation around the axis shown generates an isoprobability surface.

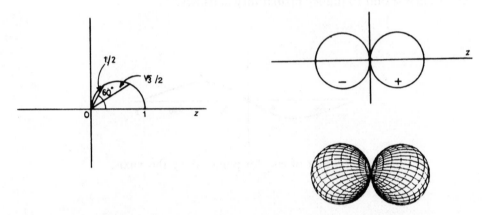

Fig. 3.12 Representation of the angular $(\cos\theta)$ of the function $\psi^2_{2p_z}$ (refs. 30–33).

A more schematic representation of a 2p orbital comes from a contour plot simulating an isoprobability surface of sufficient size such that the probability of finding the electron outside is very small (Fig. 3.11).

For an additional representation, only the angular part $Y(\theta, \phi) = \cos\theta$ is taken in the following way. For each angle θ, a point at a distance from the nucleus, equal to the value of $\cos\theta$, is chosen. All these points define two spherical surfaces which are at a tangent to each other at the position of the nucleus (Fig. 3.12).

For the equivalent representation of $Y^2(\theta, \phi) = \cos^2\theta$, see Fig. 3.13. Note, however, that the use of signs $+$ and $-$ is only applicable in the representation of $Y(\theta, \phi) = \cos\theta$.

Problem 3.10   With the same origin and in the same scale, plot $\cos\theta$ and $\cos^2\theta$ for $\theta = 0°, 30°, 45°, 60°$ and $90°$ and justify the difference of form between Figs. 3.12 and 3.13.

Although these representations do not show it directly, the sum

$$\psi^2_{2p_x} + \psi^2_{2p_y} + \psi^2_{2p_z} = \frac{1}{32\pi}(Z/a_0)^3 \rho^2 e^{-\rho} \tag{3.40}$$

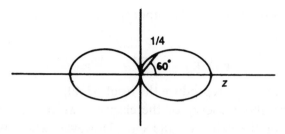

Fig. 3.13 Representation of the angular $(\cos^2\theta)$ of the function $\psi^2_{2p_z}$.

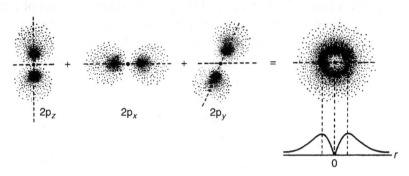

Fig. 3.14 Spherical symmetry of the sum of the probability densities for the three 2p orbitals.

is independent of the angular coordinates, that is, it has spherical symmetry. This is illustrated in Fig. 3.14.

The $2p_z$ orbital is associated with $m_\ell = 0$: a zero angular momentum around the z axis. The values $m_\ell = \pm 1$ are associated with the remaining pair of 2p orbitals but not in a one to one relation to the $2p_x$ and $2p_y$ orbitals. In fact, these functions are eigenfunctions of the hamiltonian but are not eigenfunctions of the $\hat{l}_z$ angular momentum operator (see page 27); an energy can be associated with them but not an angular momentum. The functions which are eigenfunctions of $\hat{H}$, $\hat{l}^2$ and $\hat{l}_z$ are, instead of ψ_{2p_x} and ψ_{2p_y},

$$\psi_{2p_{+1}} = f(r) \sin \theta e^{i\phi}$$
$$\psi_{2p_{-1}} = f(r) \sin \theta e^{-i\phi}. \tag{3.41}$$

These are complex wavefunctions $(e^{i\phi} = \cos \phi + i \sin \phi)$ which are related to ψ_{2p_x} and ψ_{2p_y} as follows:

$$\psi_{2p_x} = (\psi_{2p_{+1}} + \psi_{2p_{-1}})/\sqrt{2} = f(r) \sin \theta \cos \phi/\sqrt{2}$$
$$\psi_{2p_y} = -i(\psi_{2p_{+1}} - \psi_{2p_{-1}})/\sqrt{2} = f(r) \sin \theta \sin \phi/\sqrt{2} \tag{3.42}$$

($\sqrt{2}$ is a normalization factor). The appropriate probability density functions are

$$\psi_{2p_{+1}}^* \psi_{2p_{+1}} \quad \text{and} \quad \psi_{2p_{-1}}^* \psi_{2p_{-1}} \tag{3.43}$$

and these, added to $\psi_{2p_z}^2$, reproduce the result (3.40).

In discussions about energy of the electron, we can use either the pair $\psi_{2p_{+1}}$ and $\psi_{2p_{-1}}$ or the pair ψ_{2p_x} and ψ_{2p_y}. However, when dealing with the angular momentum we can only use the $\psi_{2p_{+1}}$ and $\psi_{2p_{-1}}$ functions.

Expressions (3.42) are a manifestation of the theorem discussed in Chapter 2 that a linear combination of degenerate eigenfunctions of a given operator is an eigenfunction of that operator. In the present case, $\psi_{2p_{+1}}$ and $\psi_{2p_{-1}}$ are degenerate eigenfunctions of the hamiltonian (but not of the angular momentum operator $\hat{l}_z$) and, so, ψ_{2p_x} and ψ_{2p_y} are also eigenfunctions of the same operator (but not of $\hat{l}_z$).

s, p *Hybrid orbitals*

The theorem just mentioned can also be applied to the ψ_{2s} and ψ_{2p} functions which are degenerate eigenfunctions of the hamiltonian. Therefore, any linear combination of them is also an eigenfunction of $\hat{H}$. For example,

$$\psi_{sp} = (\psi_{2s} + \psi_{2p})/\sqrt{2}$$
$$\psi'_{sp} = (\psi_{2s} - \psi_{2p})/\sqrt{2} \tag{3.44}$$

are *hybrid orbitals* made up of equal weights of ψ_{2s} and ψ_{2p}: these are called *sp hybrid orbitals*.

Problem 3.11   Confirm that the sp hybrid orbitals are orthonormal.

Other examples are:

$$\psi_{sp^2} = (\psi_{2s} + \sqrt{2}\psi_{2p})/\sqrt{3}$$
$$\psi_{sp^3} = (\psi_{2s} + \sqrt{3}\psi_{2p})/\sqrt{2} \tag{3.45}$$

which are, respectively, examples of *sp² and sp³ hybrid orbitals* (read s, p two and s, p three), where the relative weights of s and p, given by the squared coefficients, are $1:2$ and $1:3$. These hybrid orbitals will be used in some discussions of molecular structure in Chapters 4–12.

d *Orbitals*

The five 3d orbitals are all complex wavefunctions, but they can be converted into a set of real eigenfunctions of the hamiltonian by the same

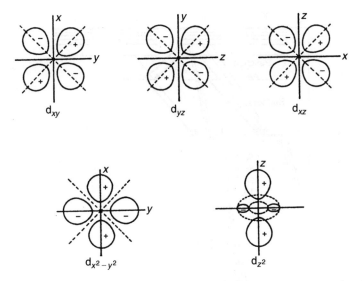

Fig. 3.15 The five 3d orbitals represented by their outer contours in the planes indicated.

procedure used for $2p_{+1}$ and $2p_{-1}$ orbitals. The contours of the real 3d orbitals are shown in Fig. 3.15 with reference to the nodal planes and to the signs of the wavefunctions.

The orbitals d_{xy}, d_{yz}, and d_{xz} all have two nodal planes, which are, respectively, the xz and yz planes, the yx and zx planes, and the xy and zy planes. The suffix xy means that the maximum amplitude occurs along the bisectors of the x and y axes and that the sign of the function at each point is the same as the sign of the product of the coordinates x and y for that point. Another orbital, $d_{x^2-y^2}$, has the same shape as the previous ones, but has maxima along the x and y axes; the sign at each point is that of the difference $x^2 - y^2$ for the coordinates of that point. The other orbital, d_{z^2}, has a maximum amplitude along the z axis, with positive values for $\pm z$ and a symmetric ring of negative values on each side of the xy plane.

Electronic transitions

The electronic spectra of monoelectronic atoms can be interpreted on the basis of transitions of the electron between orbitals involving photons. However, not all photon-induced transitions are allowed to occur in this way. Figure 3.16 shows the main transitions responsible for the emission spectrum of hydrogen.

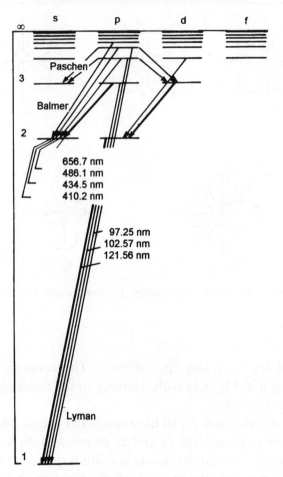

Fig. 3.16 Electron transitions involving emission of light for the H atom (thicker lines for more intense transitions).

We notice in Fig. 3.16 that there are p → s, d → p, and p → d transitions but not s → s, p → p, or d → s transitions. That is, only transitions corresponding to a change of 1 in the quantum number ℓ occur. The explanation depends upon the fact that the photon has an angular momentum (spin angular momentum) with a quantum number of 1. Thus, both in emission and absorption, conservation of the total angular momentum requires that the electron should experience a change of 1 in its orbital angular momentum. For example, when a photon is absorbed by an electron in an s orbital, the electron acquires orbital angular momentum and makes a transition to a p orbital; when absorbed by a p electron, the orbital angular momentum is increased (p → d transition) or decreased (p → s transition) depending on the relative orientations of the photon and the electron angular momenta. Photon-induced transitions involving $\Delta\ell$ different from 1 are said to be *forbidden transitions*. We say that the *selection rule* for electronic transitions in monoelectronic atoms and involving photons is

$$\Delta\ell = \pm 1.$$

In Chapter 12 we shall return to this subject.

3.3 Spin

A hydrogen atom in its ground state should have no angular momentum, because there is no net angular momentum associated with a 1s orbital ($\ell=0$). Nevertheless, it has angular momentum as can be shown through an experiment in which a beam of H atoms in a non-homogeneous magnetic field is split into two (Fig. 3.17). An experiment of this type was first performed by Stern and Gerlach in 1921, using silver atoms. These have only one valence electron (a 5s electron), like hydrogen.

These results can be explained by the electron spin, which is an intrinsic angular momentum independent of the orbital behaviour of the electron. The splitting of the incident beam is interpreted on the basis of two values for a spin magnetic quantum number, m_s. A magnetic quantum number is related to an angular momentum quantum number, a, by taking values varying between $-a$ and $+a$, in unitary jumps, that is $2a+1$ values. Thus, the observation of two values for m_s means that the spin quantum number for the electron is $s=1/2$, i.e. $m_s=\pm 1/2$.

As already mentioned (page 42), the electron spin is not included in the wave equation formulation due to Schrödinger. However, it is implicit in Dirac's formulation of quantum mechanics which associates the principles

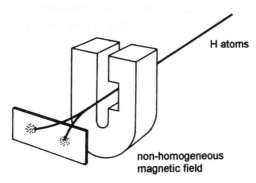

Fig. 3.17 Stern–Gerlach experiment for H atoms in their ground state.

of relativity theory. In this more complete formulation, the operators $\hat{H}, \hat{l}^2$ and $\hat{l}_z$ no longer commute, which means that stationary states (well-defined energy states) have no well-defined orbital angular momentum. We say that ℓ and m_ℓ are not good quantum numbers. However, $\hat{H}$ commutes with $\hat{j}^2$ and $\hat{j}_z$, if $\mathbf{j}$ is the correct *total angular momentum* and j_z its z component. Therefore, wavefunctions exist that are simultaneously eigenfunctions of these three operators:

$$\hat{H}\psi = E\psi$$
$$\hat{j}^2\psi = j(j+1)\hbar^2\psi \qquad (3.46)$$
$$\hat{j}_z\psi = m_j\hbar\psi$$

with quantum number j having half-integral values $(1/2, 3/2, \ldots)$ and

$$m_j = -j, -j+1, \ldots, +j.$$

The word 'total' used previously means that the correct angular momentum $\mathbf{j}$ can be thought of as the vectorial sum of an orbital angular momentum $\mathbf{l}$ with a spin angular momentum $\mathbf{s}$

$$\mathbf{j} = \mathbf{l} + \mathbf{s} \qquad (3.47)$$

with

$$j = \ell + 1/2, \quad \ell - 1/2.$$

The same can be said of the total magnetic moment of the atom in terms of the electron magnetic moments associated with the orbital and the spin angular momenta. It is thus possible to speak of a *spin–orbit* interaction, and, as a consequence, its effects on the energy values.

For each value of the total angular momentum quantum number j (within a given principal quantum number n) there is an energy level:

$$
\begin{array}{lll}
n = 1 & \ell = 0 & j = 1/2 \\
n = 2 & \ell = 0 & j = 1/2 \\
& \ell = 1 & j = 3/2, 1/2 \\
\vdots & \vdots & \vdots
\end{array}
$$

These levels are called, respectively,

$$1S_{1/2} \qquad 2S_{1/2} \qquad 2P_{3/2} \qquad 2P_{1/2} \cdots$$

(read as 'one s, one half', etc.) including the value of the principal quantum number, a reference to the value of ℓ (S for $\ell = 0$, P for $\ell = 1$, etc.) and the value of j as a subscript.

Such a spin–orbit coupling leads to a generic energy lowering and, more important, to energy splitting for $n \geq 2$. For example, for $n = 2$, there are three energy levels: $2S_{1/2}$, $2P_{1/2}$, and $2P_{3/2}$. Thus, in the emission spectrum, instead of only one $2p \rightarrow 1s$ transition (recall that the $2s \rightarrow 1s$ transition is forbidden), there will be two:

$$2P_{1/2} \rightarrow 1S_{1/2}$$
$$2P_{3/2} \rightarrow 1S_{1/2}. \tag{3.48}$$

Since this spin–orbit splitting is roughly dependent on Z^4, the effect on the spectra, although small for H atoms, is quite noticeable for sodium atoms where it was first detected. In this case, instead of one $3p \rightarrow 3s$ transition, there are two:

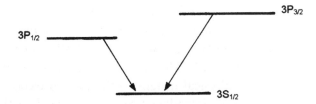

giving rise to the well-known double emission line around 590 nm.

Problem 3.12   Justify the existence of two energy values for the excited 3d[1] configuration of hydrogen.

Each relativistic wavefunction for the electron of H (and hydrogen-like atoms) has four components, as already mentioned for the simple case of a particle in a unidimensional box. For example, for the 2s orbital (level $2S_{1/2}$) we have (see, for example, ref. 34):

$$\sqrt{1/8} \begin{cases} e^{-\rho/2}(1 - \rho/2) \\ 0 \\ (1/2)i\alpha Ze^{-\rho/2}(1 - \rho/4)\cos\theta \\ (1/2)i\alpha Ze^{-\rho/2}(1 - \rho/4)e^{i\phi}\sin\theta \quad (\alpha = 1/137). \end{cases} \tag{3.49}$$

Note that the first component has a nodal surface for $\rho = 2$, but the last two are non-zero for that value of ρ. That is, the relativistic 2s orbital has no nodes. The same can be said for the other orbitals. For a discussion of relativistic effects in atom H, see refs. 35 and 36.

4

The one-electron molecule H_2^+: molecular orbitals

4.1 The wave equation and molecular orbitals

The orbital concept introduced in the previous chapter for monoelectronic atoms – the orbital as the eigenfunction of a single electron hamiltonian – can be extended to the simplest monoelectronic molecule: the H_2^+ molecular ion. The electron in H_2^+ now finds itself in the field of two nuclei, and the ground and excited state wavefunctions are no longer centred around one point in space. In H_2^+ the electron cloud surrounds both nuclei (equally) and, thus, we talk of *molecular orbitals*. This spatial distribution of electron charge around both nuclei lends physical insight to the idea of electron sharing as the basis of *covalent bonding*, introduced by G.N. Lewis in 1916 before the advent of quantum mechanics.

It is important to say, from the start, that covalent bonding and other molecular properties can be studied quantum mechanically without reference to molecular orbitals. We are referring to *valence bond theory* which, being based on the orbital concept for atoms, was, in fact, the first quantum-mechanical theory of the chemical bond. In Chapter 8, we will make a more detailed reference to this method. Our main concern in this and the coming chapters is *molecular orbital theory*.

The Schrödinger equation for H_2^+ involves three particles: two nuclei A and B of charge e and one electron (charge $-e$). Therefore, the potential energy of the system has two terms:

$$V_{nuc-nuc} = ke^2/R_{AB} \qquad (4.1)$$

$$V_{elec-nuc} = -ke^2(1/r_A + 1/r_B). \qquad (4.2)$$

The first is the contribution of the electrostatic repulsion between the two nuclei, considered as point charges at a distance R_{AB}; and the second is the contribution of the attractions between the electron and each nucleus, at variable distances r_A and r_B, respectively. Due to the vibrations that any molecule has (even at 0 K), R_{AB} is also variable. The kinetic energy of the system has an electronic component and a nuclear component associated with the molecular vibrations and rotations.

This complication due to nuclear motions is usually handled by the *Born–Oppenheimer* approximation (the American physicist J. Robert Oppenheimer (1904–1967) began his scientific career as a disciple of Born). This treats the nuclear motion and the electronic motion as entirely independent, because – in view of the big difference in mass – the velocity of the nuclei is very much less than the velocity of the much lighter electron. The nuclei are considered to be stationary in comparison with the motion of the electron. In practical terms, this means that the wave equation is solved for a fixed nuclear framework (in particular, the experimentally determined bond length) and the nuclear repulsion is then added on as a separate term. The total molecular energy becomes a sum of (approximately) independent terms: the electronic energy (kinetic plus potential), the repulsion energy (potential) between the nuclei, the molecular vibrational energy (kinetic plus potential), and the rotational and translational energy (kinetic).

If the electronic energy and the nuclear repulsion energy are calculated for different R_{AB} values and their sum is plotted as a function of the internuclear distance, a *potential energy curve* is obtained. Note, however, that the kinetic energy of the electron is part of these curves but, by application of the virial theorem, the total energy is one half of the potential energy (see page 58). Figure 4.1 shows a typical potential energy curve for a diatomic

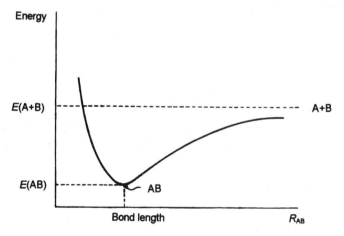

Fig. 4.1 Potential energy curve for a pair of atoms A and B, as a function of the internuclear distance, for the case of AB being a stable molecule.

molecule. The R_{AB} value for which the energy is a minimum is the bond length, an equilibrium value in view of the vibrational motion.

Within the constraints of the Born–Oppenheimer approximation, the wave equation to solve is

$$\nabla^2\psi + (2m/\hbar^2)(E - V)\psi = 0 \tag{4.3}$$

with the electronic potential energy V being given by Eq. (4.2). This equation can be solved analytically, by changing to elliptical coordinates such that

$$\mu = (r_A + r_B)/R_{AB} \qquad \nu = (r_A - r_B)/R_{AB} \tag{4.4}$$

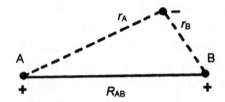

and introducing an angle ϕ as the angle made by the plane of the above triangle and a reference plane.

The wavefunction $\psi(x, y, z)$ then becomes a product of one-variable functions

$$\psi(\mu, \nu, \phi) = M(\mu)N(\nu)\Phi(\phi) \tag{4.5}$$

and the resolution of the wave equation implies the introduction of three quantum numbers, as for monoelectronic atoms. The angular part of $\psi(\mu, \nu, \phi)$ is now the simplest expression:

$$\Phi(\phi) = (1/\sqrt{2\pi})\exp(i\lambda\phi) \tag{4.6}$$

with

$$|\lambda| = 0, 1, 2, \ldots$$

being the counterpart of the quantum number m_l in atoms. It should be noted that, for $R_{AB} = 0$, H_2^+ is 'converted' into He^+, except for the nuclear mass.

Figure 4.2 shows the result of the calculation for the lower energy eigenfunctions, for variable R_{AB}. In the limit $R_{AB} = 0$, we have the energies of the orbitals 1s, 2s and 2p for He^+ (when the Z^2 dependence is considered), namely $E_{1s(He^+)} = 4E_{1s(H)}$ and $E_{2s(He^+)} = E_{2p(He^+)} = E_{1s(H)}$.

The lowest energy molecular orbital (m.o.), named 1sσ in Fig. 4.2, corresponds to an energy that is less than the value for the 1s orbital of the

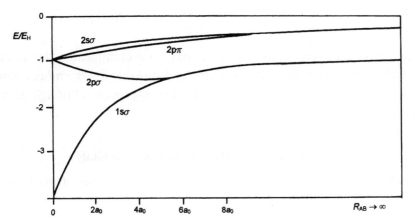

Fig. 4.2 Electronic energy of H_2^+ as a function of the internuclear distance, for the lower energy eigenfunctions. (Adapted with permission from ref. 26.)

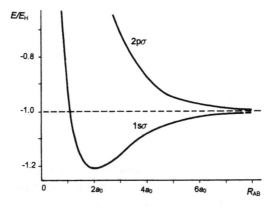

Fig. 4.3 Potential energy of H_2^+ as a function of the internuclear distance, for the two m.o.s of lowest energy.

hydrogen atom (approaching this value as $R_{AB} \to \infty$). This result is not unexpected, since the electron is attracted to two nuclei in H_2^+. It is this m.o. which describes the ground state of H_2^+. When the nuclear repulsion term is added, the potential energy curve of Fig. 4.3 is obtained. The minimum is in excellent agreement with the experimental bond length (106 pm) and with the experimental binding energy ($D_e = 2.79$ eV $= 269$ kJ mol^{-1}; see Chapter 10).

Figure 4.2 also gives the energies for the higher m.o.s. It can be seen that, when the distance R_{AB} becomes infinite, the energy of $2p\sigma$ tends to the energy value of the 1s orbital of H. Also, when $R_{AB} = 0$, the energy of $2p\sigma$

becomes equal to that of the 2p orbital of He^+. The same is true of $2p\pi$ and $2s\sigma$.

The analytical expressions for these m.o.s are complicated. In addition, this method of solving the wave equation is not viable for the more common polyelectronic molecules. That is why another approach, although approximate, is described next.

4.2 Molecular orbitals from atomic orbitals

The lowest energy m.o., $1s\sigma$, can be approximated in terms of the 1s atomic orbitals of two H atoms:

$$1s\sigma \cong \sigma_1 = N_1(\varphi_{1sA} + \varphi_{1sB}).\qquad(4.7)$$

N_1 is a normalization factor whose value is a function of the *overlap integral S* between the two orbitals $1s_A$ and $1s_B$:

$$N_1 = 1/\sqrt{2(1 + S)}$$

$$S = \int \varphi_{1sA}\varphi_{1sB}\,dv.\qquad(4.8)$$

Problem 4.1   Confirm that, with $N_1 = 1/\sqrt{2(1 + S)}$, *the integral* $\int \sigma_1^2\,dv$ *becomes unity.*

The approximation (4.7) is reasonable in view of the fact that, for points in space near nucleus A (and more distant from B), the dominant factor is the attraction to that nucleus and the wavefunction will be approximately φ_{1sA}. For points closer to B, it will be approximately φ_{1sB}.

It should be noted, however, that the same reasoning is compatible with

$$\sigma_1' = N_1'(\varphi_{1sA} - \varphi_{1sB}).\qquad(4.9)$$

Indeed, Eq. (4.9) is found to be an acceptable approximation of the second m.o., $2p\sigma$:

$$2p\sigma \cong \sigma_1' = N_1'(\varphi_{1sA} - \varphi_{1sB})\qquad(4.10)$$

with $N_1' = 1/\sqrt{2(1 - S)}$.

Problem 4.2   Confirm the expression above for the normalization factor of σ_1'.

The functions σ_1 and σ_1' are *linear combinations of atomic orbitals (l.c.a.o.)* $1s_A$ and $1s_B$. It is easily shown that they are orthogonal: thus, σ_1 and σ_1' are

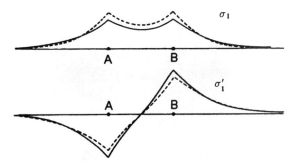

Fig. 4.4 Positive (σ_1) and negative interference (σ_1') in the internuclear region of $H_2{}^+$ (full lines: σ_1 and σ_1'; dotted lines: $1s\sigma$ and $2p\sigma$).

orthonormal functions. The plus sign in the expression of σ_1 reflects a constructive interference of the 1s atomic functions φ_{1sA} and φ_{1sB}, whereas the minus sign in the expression for σ_1' corresponds to destructive interference. In the first case, the wave amplitude increases for the region of space between the nuclei – points simultaneously close to both nuclei – whereas a decrease is obtained in the case of σ_1'. In particular, σ_1' has a node midway between the nuclei and changes sign (Fig. 4.4). It is now understood why the second m.o. was called $2p\sigma$: 2 because, for $R_{AB}=0$, it corresponds to an orbital of $n=2$ of He^+, and p because it has the symmetry of a 2p a.o. with respect to the mid-point in the bond axis. The reason for σ comes next.

4.3 Classifying molecular orbitals and electronic states

Bonding and anti-bonding σ molecular orbitals

Both σ_1 and σ_1' have cylindrical symmetry about the internuclear line; hence the designation σ, by analogy with the most symmetrical a.o.s which are the s orbitals. A σ *molecular orbital* when seen from a point on that axis looks like an s orbital. The function σ_1 is always positive and is symmetric with respect to inversion at the molecular centre, whereas σ_1' changes sign at this point. It is said that they have different *parity*: σ_1 and σ_1' are represented, respectively, by σ_g and σ_u, the indices g and u coming from the german words 'gerade' and 'ungerade' for 'even' and 'odd', respectively.

The probability densities for σ_1 and σ_1' are expressed as follows

$$\sigma_1^2 = N_1^2(\varphi_{1sA}^2 + \varphi_{1sB}^2 + 2\varphi_{1sA}\varphi_{1sB}) \tag{4.11}$$

$$\sigma_1'^2 = N_1'^2(\varphi_{1sA}^2 + \varphi_{1sB}^2 - 2\varphi_{1sA}\varphi_{1sB}) \tag{4.12}$$

and are semi-quantitatively represented in Fig. 4.5 for points on the internuclear line.

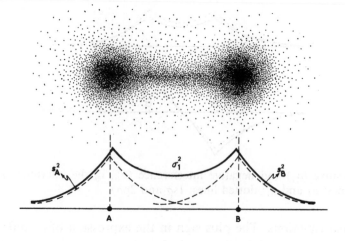

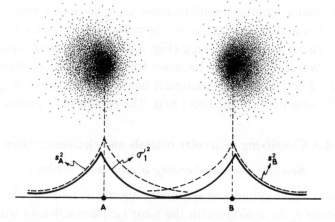

Fig. 4.5 Comparison between the probability densities for σ_1 and σ_1' of H_2^+ for points on the internuclear line (the effect of the difference in the normalization factors is ignored).

With σ_1^2, there is an increase of probability density $2N_1^2\varphi_{1sA}\varphi_{1sB}$ in the internuclear region – where both atomic functions have significant values – whereas, with $\sigma_1'^2$, there is a decrease $2N_1'^2\varphi_{1sA}\varphi_{1sB}$. This migration of charge in opposite directions – towards the internuclear region in the case of σ_g, and towards the outside in the case of σ_u – is consistent with a *bonding effect* for an electron in σ_g and an *anti-bonding effect* for an electron if described by σ_u. In fact, an electron in the internuclear region attracts both nuclei towards each other, thus opposing the nuclear repulsion forces, whereas

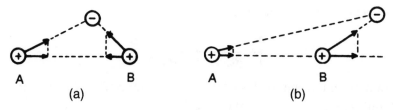

Fig. 4.6 Forces acting on the nuclei of H_2^+ due to the electron in different positions: (a) bonding effect; (b) anti-bonding effect.

outside it contributes to the separation of the nuclei (Fig. 4.6). Therefore, σ_g is a *bonding molecular orbital* and σ_u is an *anti-bonding molecular orbital*.

Figure 4.7 shows the isoprobability contour plots for both m.o.s in a plane containing the nuclei. Computer programs are available that readily calculate and plot electron density and other properties of m.o.s (see ref. 37 and Chapters 6 and 7).

According to the above simplistic considerations, it is expected that the energy of the electron in σ_g is less than in σ_u. In fact, a negative charge

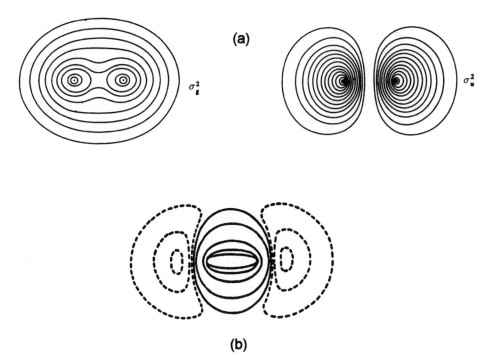

Fig. 4.7 (a) Isoprobability contour plots in a plane containing the nuclei of H_2^+ for σ_g and σ_u, and (b) density difference map for σ_g^2 compared to $(\varphi_{1sA}^2 + \varphi_{1sB}^2)/2$ (full lines for electron gain and dotted lines for electron loss).

mid-way between two positive charges

has a lower potential energy than if it lies at a distance $R_{AB}/2$ from one positive charge and $R_{AB} + R_{AB}/2$ from the other:

However, energy considerations are more subtle, as shown by Ruedenberg and co-workers (ref. 38 and references therein) and clearly discussed in refs. 39 and 40. One of the main criticisms is that the above model completely neglects the significance of electronic kinetic energy. If we want to explain why the energy of the electron in H_2^+ is less than in the H atom (both species in their ground states) then we must start by considering that its kinetic energy decreases as a result of an extension in space of the electron to a larger volume (see expression 2.69). Also, if migration of charge to the region between the nuclei leads to a decrease of the potential energy, the greater the electron density between the nuclei the less it is near each nucleus: an increased potential energy. However, there is another effect of this charge depletion near the nuclei: the electron cloud contracts and the potential energy may decrease after all. At the same time, because of this contraction, the kinetic energy does not decrease as much and may even increase (especially in the directions perpendicular to the internuclear axis). In fact, according to the virial theorem, the potential energy decreases on bond formation by an amount which is twice the increase in the electron kinetic energy. Thus, orbital contraction plays an important role in stabilization by bond formation.

Expressions for the energies associated with the σ_1 and σ_1' m.o.s can be obtained through the following integrals (expectation value of the hamiltonian H):

$$E_1 = \langle \sigma_1 | \hat{H} | \sigma_1 \rangle$$
$$E_1' = \langle \sigma_1' | \hat{H} | \sigma_1' \rangle. \tag{4.13}$$

Replacing σ_1 by Eq. (4.7) and representing φ_{1sA} by s_A and φ_{1sB} by s_B, we have

$$E_1 = \langle \sigma_1 | \hat{H} | \sigma_1 \rangle = \langle s_A + s_B | \hat{H} | s_A + s_B \rangle / [2(1+S)]$$
$$= (\hat{H}_{aa} + \hat{H}_{ab} + \hat{H}_{bb} + \hat{H}_{ba}) / [2(1+S)]$$
$$= (\alpha + \beta) / (1+S) \tag{4.14}$$

with

$$H_{aa} = \langle s_A|\hat{H}|s_A\rangle = \alpha \qquad H_{bb} = \langle s_B|\hat{H}|s_B\rangle = \alpha$$
$$H_{ab} = \langle s_A|\hat{H}|s_B\rangle = \beta \qquad H_{ba} = \langle s_B|\hat{H}|s_A\rangle = \beta. \tag{4.15}$$

Since, for $R_{AB} \to \infty$,

$$\alpha \to E_{1sH}$$
$$\beta \to 0$$

and, for $R_{AB} \to 0$,

$$\alpha = \beta,$$

we conclude that α and β are negative quantities. They are called, respectively, the *Coulomb integral* and *resonance integral*. The first, the expectation value of $\hat{H}$ for the 1s function of H, is in the limit the energy of the hydrogen 1s orbital. The resonance integral is related to the energy of the electron density $s_A s_B$, especially significant in the internuclear region.

For σ_1', the energy is

$$E_1' = (\alpha - \beta)/(1 - S) \tag{4.16}$$

a value higher than E_1, as $\beta < 0$ (and S not too close to 1).

Problem 4.3   Confirm the expression (4.16).

We have, thus, the following energy diagram:

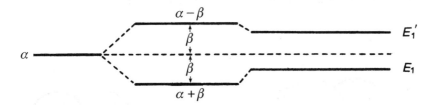

This diagram shows the effect of the difference in the normalization factors. For $S = 0.60$, a value adequate for the internuclear distance of 106 pm in H_2^+, we calculate $N_1 = 0.56$ and $N_1' = 1.10$. As a consequence, the anti-bonding m.o. σ_1' is stabilized relative to $\alpha - \beta$, whereas the bonding m.o. σ_1 is destabilized relative to $\alpha + \beta$.

Problem 4.4   Find that, with $\alpha = -14$ eV and $\beta = -10$ eV, the anti-bonding m.o. is stabilized to a lesser extent than the bonding m.o. is destabilized (i.e. the bonding m.o. is less bonding than the anti-bonding m.o. is anti-bonding).

A calculation of the energy for variable internuclear distances (including atomic orbital contraction) leads, after adding the nuclear repulsion energy, to a potential energy curve similar to Fig. 4.1, with the minimum at 106 pm, in exact agreement with experiment, and producing a value for the dissociation energy which differs by about 20% from the experimental value.

Problem 4.5   The closer the nuclei in $H_2{}^+$ *the larger is S, the more negative is* β *and the smaller is the normalization factor* N_1. *On the other hand, the smaller* R_{AB}, *the smaller is the internuclear region for the bonding effect and the smaller the region for electron delocalization. Based on these considerations, criticize the sentence 'The greater the overlap integral the stronger the bond'.*

Bonding and anti-bonding π molecular orbitals

We now turn to the excited m.o.s $2s\sigma$ and $2p\pi$ of Fig. 4.2 which become a.o.s of H ($n=2$) when the internuclear distance goes to infinity. The $2p\pi$ m.o. is doubly degenerate and can be approximated as follows

$$2p\pi \cong \pi_y = N_2(y_A + y_B)$$

$$\cong \pi_z = N_2(z_A + z_B) \tag{4.17}$$

where y and z represent 2p atomic orbitals of H (A and B). The internuclear axis was chosen as the x axis. Each one of the linear combinations (Eq. (4.17)) involves 2p orbitals with parallel axes (Fig. 4.8).

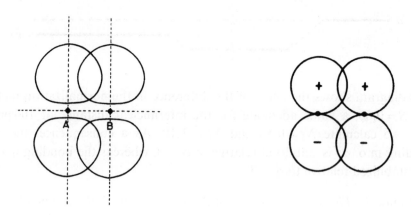

Fig. 4.8 Two representations of a pair of 2p orbitals with parallel axes (see page 62).

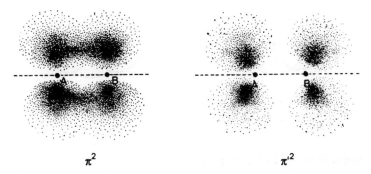

$$\pi^2 \qquad\qquad\qquad \pi'^2$$

Fig. 4.9 Electron cloud representations of the electron density for bonding and anti-bonding π molecular orbitals.

In Eq. (4.17) there is constructive interference of the two a.o.s involved in each m.o. which means that these m.o.s are bonding, although of higher energy than σ_1 and σ_1'. For π_y (and for each $2p_y$ orbital) the xz plane is a nodal plane, and for π_z (and for $2p_z$) the nodal plane is xy. On account of this, they are called π *molecular orbitals*; when viewed from the direction of the x axis, they resemble p atomic orbitals. It is noted that they change sign upon inversion in the centre of the molecule; the parity is u (π_u).

Destructive interference would lead to anti-bonding π molecular orbitals:

$$\pi_y' = N_2'(y_A - y_B)$$
$$\pi_z' = N_2'(z_A - z_B). \qquad (4.18)$$

Problem 4.6   Show that the m.o.s (4.18) have parity g.

Problem 4.7   Compare the energies of π_u and π_g as a function of the integrals α and β (see page 79).

Figure 4.9 shows, in a semi-quantitative manner, the electron clouds corresponding to the electron density for a bonding and an anti-bonding π molecular orbital.

Other molecular orbitals

Other excited m.o.s can be constructed from the 2s and $2p_x$ atomic orbitals of A and B. The previous examples already point to a general principle of conservation of the number of orbitals: from two a.o.s two m.o.s can be constructed – one bonding, one anti-bonding – from four a.o.s four m.o.s can be constructed, etc. Considering Fig. 4.10, the four m.o.s constructed from the 2s and $2p_x$ functions of A and B are σ m.o.s.

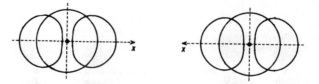

Fig. 4.10 The 2s and $2p_x$ atomic orbitals appropriate to the construction of four excited σ molecular orbitals for H_2^+.

The general form of those m.o.s is

$$a(\varphi_{2sA} + \varphi_{2sB}) + b(\varphi_{2xA} + \varphi_{2xB}) = (a\varphi_{2sA} + b\varphi_{2xA}) + (a\varphi_{2sB} + b\varphi_{2xB})$$
$$(4.19)$$

where the numerical coefficients are different for 2s and 2p orbitals but are the same for A and B because of symmetry. It is not correct to combine the 2s and the $2p_x$ atomic orbitals separately, because there is also a net overlap between orbitals of the two types. The various combinations are graphically shown next:

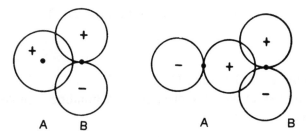

Fig. 4.11 Orbitals of A and B which are orthogonal.

The contributions from 2s and $2p_x$ of the same atom can be associated in Eq. (4.19) in mixed atomic functions which have already been introduced on page 64: *hybrid orbitals*. An s, p hybrid orbital has two lobes and a nodal surface like a p orbital, but one lobe is much larger resembling an s orbital.

From 3s, 3p and 3d a.o.s other excited m.o.s can be defined: σ, π and δ molecular orbitals. There are, however, some combinations that are not considered. For example, we have not considered cases like

$$1s_A \pm 2y_B \qquad 1s_A \pm 2z_B \qquad 2x_A \pm 2y_B. \qquad (4.20)$$

The reason is that these combinations involve orthogonal atomic orbitals, as illustrated in Fig. 4.11. If the overlap integral is zero, the resonance integral is also zero. In such a case, there is no way of defining two m.o.s: one of smaller energy, one of larger energy. For example, it can be seen in Fig. 4.11 that the constructive interference of the 1s orbital of A with the positive lobe of the $2p_y$ orbital of B is cancelled by the destructive interference with the negative lobe.

Electronic states

As we have seen in the case of the H atom, the electronic states of H_2^+ are affected by the spin–orbit interaction and can be designated according to the electron angular momentum about the internuclear axis.

In the configuration of lowest energy, the electron occupies a m.o. of cylindrical symmetry, σ_g, and the corresponding angular momentum is zero: $\lambda = 0$ in expression (4.6). This electronic state is labelled Σ_g, and a $+$ sign is added as a superscript to indicate that the wavefunction keeps its sign upon reflection in a plane containing the internuclear axis (symmetry operation σ_v). The spin angular momentum is also shown, through the spin multiplicity $2s + 1$ (the number of m_s values possible), in this case, 2 ($s = 1/2$).

This information is shown as a superscript on the left. Thus:

$$^2\Sigma_g^+$$

for the electronic ground state of $H_2{}^+$.

For an excited state with the electron in a bonding π molecular orbital, the label is $^2\Pi_u$.

5

Many-electron atoms and the orbital concept

5.1 Wavefunction and the Pauli principle

Our main concern in the study of many-electron atoms (and many-electron molecules) continues to be the energy eigenproblem. Accordingly, the wavefunctions of interest are those that are eigenfunctions of the hamiltonian.

The helium atom ($Z=2$), with two electrons, has just one mono-ionization energy, hence the two electrons have the same energy. The lithium atom ($Z=3$) and the beryllium atom ($Z=4$) both have two ionization potentials, but B ($Z=5$), C ($Z=6$), ..., Ne ($Z=10$) have three. For Na ($Z=11$) there are four. We are referring to the process

$$X \rightarrow X^+ + e^- \tag{5.1}$$

the ΔU value of which depends on the level of energy from where the electron is removed, and not to the subsequent ionization processes. We shall see later that the ionization potentials can be used to obtain only approximate electronic energies (except in one-electron atoms for which the exact energy is the symmetric of the ionization potential), but the number of them is a clear indication of the number of electronic energy values to consider. It is thus possible to establish the electron distributions in Fig. 5.1.

Any of the distributions shown in Fig. 5.1 can be reproduced by assuming atomic orbitals, by analogy with the H atom, providing two conditions are satisfied:

(a) a maximum of two electrons for each orbital;
(b) the energy of the 2s orbital is lower than the 2p.

Condition (a) is not obvious, but condition (b) is not entirely unexpected because electron repulsions will be different for 2s and 2p electrons, in view of the different shapes of the corresponding orbitals. Condition (a) was first

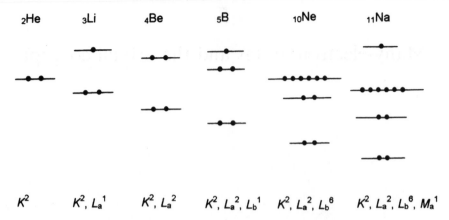

Fig. 5.1 Electron distribution among energy levels $K, L, M, \ldots$ (not to scale).

introduced by Pauli in 1924 (the *exclusion principle*) to explain the absence of certain lines in the spectrum of helium. The *Pauli principle* also acknowledges that the two electrons in a given orbital have opposite spins, that is, the respective angular momenta (and the magnetic moments associated with them) cancel each other. For the two electrons of He in the ground state, both the total spin angular momentum and the total orbital angular momentum are zero, hence the total angular momentum vanishes.

By analogy with the H atom, we can write for Na, for example, the following *electronic configuration*

$$1s^2, 2s^2, 2p^6, 3s^1 \tag{5.2}$$

where the six 2p electrons completely fill the three degenerate orbitals and where use is made of the following sequence of orbital energies:

$$E_{1s} < E_{2s} < E_{2p} < E_{3s} < \cdots \tag{5.3}$$

The electrons of each pair in orbitals 1s, 2s and 2p have opposite spins. The 3s electron is called a *valence-shell electron*, and the 10 electrons associated with lower values of the quantum number n are *core electrons*.

It is clear that what fundamentally distinguishes the study of a polyelectronic atom from a monoelectronic one is the Pauli principle and the repulsion between electrons. The Pauli principle is directly related to the spin of the electron, whose quantum number is 1/2: the electron is called a fermion particle. Let us first consider this question in more detail. For such a purpose, we will ignore the electron repulsion, for the moment. This means that each electron behaves independently of the others, as if it were the only one in the atom. The atomic orbitals are then exactly the same as those for the corresponding hydrogen-like atom. For example, each of the two electrons of He would occupy a 1s orbital identical to that of He^+.

In these circumstances, the hamiltonian for the two electrons of He is

$$\hat{H}(1,2) = \hat{H}(1) + \hat{H}(2) \qquad (5.4)$$

where $\hat{H}(i)$ includes the kinetic energy operator for electron i and the potential energy for the attraction between the nucleus and the electron. Analogously, the total energy is the sum of the individual energies

$$E = E_1 + E_2 \qquad (5.5)$$

and the wavefunction for the ground state is

$$\psi(1,2) = \varphi_{1s}(1)\varphi_{1s}(2) \qquad (5.6)$$

with

$$\psi^2(1,2) = \varphi_{1s}^2(1)\varphi_{1s}^2(2). \qquad (5.7)$$

To express the wavefunction as a product of functions, one for each electron, is in agreement with the fact that, in the absence of interaction between the electrons, the probability of finding electron 1 in point (x_a, y_a, z_a) and electron 2 in point (x_b, y_b, z_c) is the product of the individual probabilities $\varphi_{1s}^2(x_a, y_a, z_a)$ and $\varphi_{1s}^2(x_b, y_b, z_c)$. This is the result of the joint probability for two independent events being simply the product of the individual probabilities:

$$P(1,2) = P(1)P(2). \qquad (5.8)$$

Since the two electrons of He in the configuration $1s^2$ have opposite spins, the wavefunction should include this information. For example,

$$\psi(1,2) = \varphi_{1s}(1, \alpha)\varphi_{1s}(2, \beta) \qquad (5.9)$$

which can also be represented as

$$\psi(1,2) = \varphi_{1s}(1)\bar{\varphi}_{1s}(2) \qquad (5.10)$$

where, conventionally, a stroke over φ means β spin. The atomic functions $\varphi_{1s}(1, \alpha)$ and $\varphi_{1s}(2, \beta)$ are called *spin-orbitals*.

However, since the electrons are indistinguishable, it is not possible to say that it is electron 1 that has α spin whereas electron 2 has β spin. Thus, the function

$$\psi(1,2) = \bar{\varphi}_{1s}(1)\varphi_{1s}(2) \tag{5.11}$$

would be equally acceptable. The wavefunction must be a combination of Eqs. (5.10) and (5.11) with equal weights. There are two possibilities:

$$\psi^s(1,2) = \varphi_{1s}(1)\bar{\varphi}_{1s}(2) + \bar{\varphi}_{1s}(1)\varphi_{1s}(2) \tag{5.12}$$

and

$$\psi^a(1,2) = \varphi_{1s}(1)\bar{\varphi}_{1s}(2) - \bar{\varphi}_{1s}(1)\varphi_{1s}(2) \tag{5.13}$$

(ignoring the normalization factors). The possibility of a combination involving a minus sign is justified because physical meaning can only be attributed to squared wavefunctions. The suffix 's' refers to a symmetric function, that is, one which does not change sign (let alone absolute value) when the indices 1 and 2 specifying the electrons are interchanged, which is the case for ψ^a: an antisymmetric function.

Both functions conform to the need that the electron density ψ^2 must be unchanged if two electrons are exchanged, because they are identical. It is found, however, that only ψ^a is acceptable, being the only one satisfying the Pauli principle. In fact, the function becomes zero, $\psi^a(1,2) \equiv 0$ (a situation to be excluded from reality) if the two electrons are assigned the same spin (α or β). This conclusion, that the wavefunction must be antisymmetric with respect to the interchange of the labels of any two electrons, is a requirement for any polyelectronic atom (or molecule). This is called the *generalized Pauli principle*. Indeed, it is found universally in nature that particles of spin 1/2, such as electrons, behave such that an exchange of identical particles (fermions) causes the wavefunction to change sign.

It should be noted that the exchange of electrons is not a physical phenomenon but a correction to any initial description which wrongly assumes that the electrons can be distinguished from each other. Since indistinguishable particles by definition cannot be labelled, electron exchange in wavefunctions should be regarded as an exchange of the electron coordinates (ref. 10).

It is easy to write down the form of an antisymmetric wavefunction for a polyelectronic atom if we recognize that Eq. (5.13) can be written in the form of a determinant (*Slater determinant*) which is (again ignoring the normalization factor)

$$\psi^a(1,2) = \begin{vmatrix} \varphi_{1s}(1) & \bar{\varphi}_{1s}(1) \\ \varphi_{1s}(2) & \bar{\varphi}_{1s}(2) \end{vmatrix} = \begin{vmatrix} \varphi_{1s}(1) & \varphi_{1s}(2) \\ \bar{\varphi}_{1s}(1) & \bar{\varphi}_{1s}(2) \end{vmatrix} \tag{5.14}$$

a form which can easily be generalized. For example, for Be in the configuration $1s^2$, $2s^2$, we can write (ignoring the normalization factor):

$$\psi(1,2,3,4) = \begin{vmatrix} \varphi_{1s}(1) & \varphi_{1s}(2) & \varphi_{1s}(3) & \varphi_{1s}(4) \\ \bar{\varphi}_{1s}(1) & \bar{\varphi}_{1s}(2) & \bar{\varphi}_{1s}(3) & \bar{\varphi}_{1s}(4) \\ \varphi_{2s}(1) & \varphi_{2s}(2) & \varphi_{2s}(3) & \varphi_{2s}(4) \\ \bar{\varphi}_{2s}(1) & \bar{\varphi}_{2s}(2) & \bar{\varphi}_{2s}(3) & \bar{\varphi}_{2s}(4) \end{vmatrix}. \tag{5.15}$$

This function acknowledges the meaning of $1s^2$, $2s^2$: two (non-specified) electrons are 1s (with opposite spins) and two (non-specified) are 2s. That Eq. (5.15) is antisymmetric is easily confirmed upon realizing that to exchange the indices of two electrons is equivalent to interchanging two columns of the determinant, which changes sign as a consequence. Expression (5.15) is also in full agreement with the exclusion principle as a result of the property of determinants which states that a determinant vanishes if two lines are equal (this is the case when the electrons in one of the orbitals are 'forced' to have the same spin).

He, Be and Ne are examples of atoms with complete orbitals (*closed-shell configurations*). For atoms with incomplete orbitals (*open-shell configurations*) more than one Slater determinant is necessary.

Problem 5.1 *Write out the two possible Slater determinants for the electron configuration $1s^2$, $2s^1$ of lithium.*

There is another property of determinants which is important to consider: determinants remain unchanged when the respective elements are subject to some specific operations (*a unitary transformation*) as can easily be illustrated in Problem 5.2.

Problem 5.2 *Confirm that*

$$\begin{vmatrix} a & b \\ c & d \end{vmatrix} = \begin{vmatrix} a & b + \lambda a \\ c & d + \lambda c \end{vmatrix}$$

for any value of λ.

For example, the wavefunction (5.15) is not changed if we replace φ_{1s} by $\chi_1 = \varphi_{1s} + \lambda\varphi_{2s}$, that is, if we replace any row of the determinant by a linear combination of this row with other rows. Such mathematical alterations 'inside' the determinant do not correspond to any physical change, because the wavefunction ψ remains unchanged. Therefore, atomic orbitals truly only have a mathematical meaning in so far as they are the basis of the wavefunction ψ expressed as a Slater determinant (however, see Section 5.4). It is this wavefunction which has physical meaning through $\psi^*\psi$ in the calculation of expectation values:

$$\langle A \rangle = \langle \psi | \hat{A} | \psi \rangle. \tag{5.16}$$

5.2 Electron repulsion: orbitals, an approximation

Besides being antisymmetric, the wavefunction for a polyelectronic atom must also reflect the ever present electron–electron repulsion. In Section 5.1, the electron repulsion was ignored in order to facilitate the analysis of the Pauli principle. However, huge errors can be produced when electron repulsion is neglected. For example, the ionization potential so calculated for He would be 5250 kJ mol^{-1} (4 hartree), which is the value for He$^+$, but the experimental value is 2372 kJ mol^{-1}.

The interelectronic repulsions are represented in the wave equation by the corresponding potential energy term

$$V_{\text{elec-elec}} = ke^2 \sum_{i<j}(1/r_{ij}) \tag{5.17}$$

where r_{ij} is the distance (variable) between electron i and electron j. The summation is for $i<j$ to warrant that, after taking $(1/r_{12})$ for the repulsion between electron 1 and electron 2, $(1/r_{21})$ which would correspond to the same repulsion is not considered. The presence of $(1/r_{ij})$ terms in the Schrödinger equation

$$-(\hbar^2/2m)\sum_{i}(\partial^2/\partial x_i^2 + \partial^2/\partial y_i^2 + \partial^2/\partial z_i^2)\psi(1,2,\ldots,i,\ldots)$$
$$+ V\psi(1,2,\ldots,i,\ldots) = E\psi(1,2,\ldots,i,\ldots) \tag{5.18}$$

with

$$V = k(-Ze^2/r_1 - Ze^2/r_2 - \cdots + e^2/r_{12} + e^2/r_{13} + \cdots) \tag{5.19}$$

prevents an analytical solution of the equation. On the other hand, there cannot be one function for each electron, because the description of the behaviour of one electron is not independent of the description of the behaviour of another. Truly, one could say that there are no orbitals in polyelectronic systems, since the concept of orbital is tied up with each single electron (ref. 41). However, the fact that electronic configurations can be qualitatively reproduced by using an orbital framework imported from monoelectronic atoms suggests that the orbital concept still plays a role in polyelectronic atoms (ref. 42). But, in view of the errors committed when ignoring electron repulsion altogether, such a role must take this repulsion into account.

The way out lies in the following approximation: each electron, as far as the calculation is concerned, is considered as interacting with a smoothed-out averaged density of the other electrons. For example, the description of the 11th electron of Na – electron 3s in Eq. (5.2) – is based on an average field due to the remaining 10 electrons (besides the nucleus attraction). In this way, a central field is simulated and one-electron functions, i.e. orbitals, reappear, this time as an approximation.

There is an obvious vicious circle in this approach: if the spatial distribution of each electron is one of the unknowns, how can we speak of averaged distributions? The answer is an iterative numerical calculation as demonstrated originally by the British physicist Douglas Hartree in 1928. In 1930, the method was improved by the Russian physicist Vladimir Fock who adapted the method to antisymmetric wavefunctions as required by the Pauli principle. The *Hartree–Fock method* is a numerical calculation that can be summarized in the following steps;

(a) Begin with a set of orthonormal hydrogen-like orbitals (or some others already adequately modified) φ_i, for example $1s^2$, $2s^2$, $2p^6$ for Ne.
(b) Calculate the averaged repulsion energy of each electron i by the distribution $\varphi_j^* \varphi_j$ of each of the others

$$\int \varphi_j^* \varphi_j (1/r_{ij}) \, dv \qquad (5.20)$$

and obtain a value for the total potential energy of electron i:

$$V_i = -kZe^2/r_i + ke^2 \sum_j \left[\int \varphi_j^* \varphi_j (1/r_{ij}) \, dv \right]. \qquad (5.21)$$

(c) Make use of this potential in the calculation of a new set of atomic orbitals and a new energy.

(i) Should the electron exchange due to their indistinguishability be ignored (as in Hartree's original work), the wavefunction would be a single product of occupied spin-orbitals and it would suffice to numerically solve the individual wave equations

$$(-\hbar^2/2m)\nabla_i^2\varphi_i + V_i\varphi_i = E_i\varphi_i \tag{5.22}$$

to obtain the next set of improved atomic orbitals.

(ii) However, the determinantal form of the wavefunction, as in Eq. (5.15), introduces integrals that differ from the terms $V_i\varphi_i$ by a permutation of indices i and j:

$$-ke^2 \sum_j \left[\int \varphi_j^*\varphi_i(1/r_{ij})\,dv \right]\varphi_j. \tag{5.23}$$

These terms should also be calculated and included in the determination of the new set of orbitals and the new value for the energy. They can be seen as a new potential (although not corresponding to any new force or to a multiplicative operator like V_i) called an *exchange potential*.

(d) Repeat procedures (b) and (c) using this new set of functions and continue the cyclic process (always keeping the functions orthonormalized) until the set of orbitals obtained in a given cycle is practically identical to the previous one used as a starting basis. Consistency is reached between the field the set of orbitals produces and the field which contains their origin. The orbitals obtained are called *SCF Hartree–Fock atomic orbitals*, where SCF stands for 'self-consistent field'.

(e) Meanwhile, the total electronic energy calculated for the system, that is, the mean value of the hamiltonian with respect to the wavefunction ψ obtained after each cycle has converged to a lower limit. According to a basic principle in quantum mechanics, called the *variation principle*, this means that the final function ψ is the best approximate wavefunction within the framework of the averaged electron repulsion. This principle states that the best approximate eigenfunction of the hamiltonian is that which leads to a minimum expectation value of the hamiltonian $\langle\psi|\hat{H}|\psi\rangle$. That is, an approximate energy calculated as the expectation value of the hamiltonian for an approximate wavefunction is always greater than the exact energy value.

SCF HF orbitals are numerical functions, that is a collection of numbers for various coordinates in space around the nucleus. They can of course be plotted graphically. The angular dependence $Y(\theta,\phi)$ is the same as that of the corresponding hydrogen-like orbitals. That is why we can still refer to s, p, d, etc. orbitals. However, the radial part, $R(r)$, is different. This is the reason why differences occur between the energies of orbitals with the same

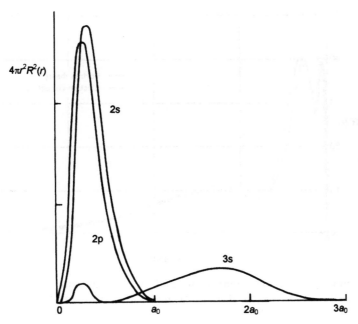

Fig. 5.2 Radial distributions $4\pi r^2 R^2(r)$ for the 2s, 2p and 3s SCF HF orbitals of Na.

principal quantum number n. Figure 5.2 shows the radial distributions $4\pi r^2 R^2(r)$ for the 2s, 2p and 3s SCF HF orbitals of Na.

Problem 5.3   The calculated total radial distribution $4\pi r^2\psi^2$ for the helium atom, obtained by two methods: the SCF method and ignoring electron repulsion (ref. 13) is shown. Interpret.

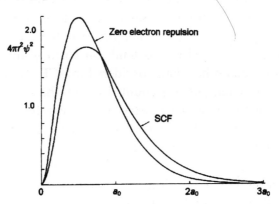

At the time of Hartree and Fock, there were no computers and the calculations would take a long time. Nowadays, sophisticated calculations can be carried out easily (see, for example, refs. 13 and 43).

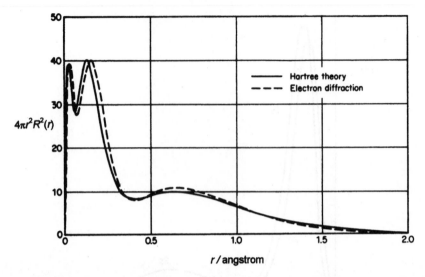

Fig. 5.3 Comparison of the experimental and SCF radial distributions of the total electron density for the argon atom (reprinted with permission from ref. 44; copyright 1953, American Physics Society).

The SCF HF orbitals provide a solid ground for the electronic configurations of atoms, enable a rationalization of the Periodic Table, and are, in general, in good agreement with experimental data (Fig. 5.3). For some purposes, however, an important source of error associated with the use of SCF atomic orbitals is that averaged electron repulsions do not represent the true situation. In reality, for the helium atom for example, the two electrons will have their instantaneous positions correlated, there being a higher probability of the two electrons being on opposite sides of the nucleus than there is of them being both on the same side, which is not reproduced by SCF theory. This *Coulomb correlation* between the electrons leads to a lowering of the energy, as electrons tend, on electrostatic grounds, to be farther away from each other than considered in SCF theory. A fraction of this error in SCF calculations is accounted for by the so-called *Pauli correlation* associated with the antisymmetry of the wavefunction: electrons of the same spin tend to be farther apart than electrons of opposite spins.

As a result of the Pauli correlation, the effective Coulomb interaction between a 2s electron of spin α, say, and a 1sα electron will be different from the interaction with a 1sβ electron. Therefore, the 1s spin-orbitals will have slightly different spatial parts. This is considered in the so-called *unrestricted Hartree–Fock* calculations.

The variation principle enables a partial correction of the SCF wavefunction by simulating the Coulomb correlation by a mixture of the fundamental configuration wavefunction with excited configuration functions, the mixing coefficients being determined subject to the condition of minimizing the expectation value of the hamiltonian. This method is known as *configuration interaction*. Another method which deals successfully with this problem is the *density-functional theory*. In this theory the energy of the ground state is directly related to the electron density, and not to the wavefunction, in an iterative mathematical process that has the additional advantage of being much less demanding than the Hartree–Fock and configuration interaction calculations. The orbital concept can be retained in density-functional theory, as one-electron functions which add up to yield the total electron density; however, they lack any physical meaning related to one-electron energies (see Section 5.7).

5.3 Total electronic energy

From the expression of the wavefunction for a closed-shell atom in terms of doubly occupied atomic orbitals it can be shown that the total probability density is given by the sum of the contributions of each occupied orbital:

$$\psi^*\psi = 2\sum_{i}^{\text{occ.}} \varphi_i^*\varphi_i. \tag{5.24}$$

Since the sums of the electron densities of three np orbitals, five nd orbitals, etc. all have spherical symmetry, as do the s orbitals, the total electron density Eq. (5.24) is spherically symmetric and isolated atoms are spherical. This conclusion is not altered by including corrections regarding the electron correlation.

An additivity similar to Eq. (5.24) does not apply to the total energy. Because of electron repulsion and electron exchange, the total electronic energy is *not* equal to the sum of the energies of the electrons. So, if ε_i is the energy associated with each (doubly) occupied orbital,

$$E \neq 2\sum_{i}^{\text{occ.}} \varepsilon_i. \tag{5.25}$$

It is noted that the energy of electron 1 includes the energy of repulsion between electron 1 and electron 2, and the energy of electron 2 includes the energy of repulsion between electron 2 and electron 1. Therefore, when summing the energies of 1 and 2, the same repulsion $1 \Leftrightarrow 2$ is counted twice. The same can be said for the exchange potential: the exchange energy is counted twice. The correct expression is obtained by subtracting, once, all the repulsion energy terms and all the exchange energy terms from the right hand side of Eq. (5.25).

Each electron in an orbital of energy ε_i suffers repulsion from the two electrons in orbital φ_j. This is a positive contribution to the energy: $2J_{ij}$. Each electron of φ_i exchanges with the electron of same spin in φ_j. This is a negative (stabilization) contribution to the energy: $-K_{ij}$. Therefore, the total electronic energy E is

$$E = \sum_i^{\text{occ.}} \left[2\varepsilon_i - \sum_j^{\text{occ.}} (2J_{ij} - K_{ij}) \right]. \tag{5.26}$$

The repulsion term J_{ij} is the classical Coulomb repulsion between the electron with charge cloud $\varphi_i^* \varphi_i$ and each of the two electrons with charge clouds $\varphi_j^* \varphi_j$. The integral is (ignoring some constants)

$$J_{ij} = \langle \varphi_i(1)\varphi_j(2)|(1/r_{12})|\varphi_i(1)\varphi_j(2)\rangle. \tag{5.27}$$

The exchange term differs by one permutation of the indices 1 and 2 for the electrons:

$$K_{ij} = \langle \varphi_i(1)\varphi_j(2)|(1/r_{12})|\varphi_j(1)\varphi_i(2)\rangle. \tag{5.28}$$

Also, the energy ε_i of each m.o. can be expressed as the sum of a term ε_i^0 representing the energy an electron in φ_i would have if it were the only electron in the atom (kinetic energy plus nuclear attraction terms) with repulsion and exchange energies:

$$\varepsilon_i = \varepsilon_i^0 + \sum_j^{\text{occ.}} (2J_{ij} - K_{ij}). \tag{5.29}$$

Substituting the value of ε_i of Eq. (5.29) into Eq. (5.26), we obtain the alternative expression

$$E = \sum_i^{\text{occ.}} \left[2\varepsilon_i^0 + \sum_j^{\text{occ.}} (2J_{ij} - K_{ij}) \right]. \tag{5.30}$$

A physical interpretation of the exchange energy associated with the Pauli principle can be advanced (see, for example, ref. 45). The spin correlation introduced by the antisymmetrization of the wavefunction means that, independently of the Coulomb repulsion, the probability of two electrons being close in space is smaller for electrons with the same spin than for opposite spins. Because of this diminished probability, we can imagine each electron as if it was accompanied by a zone of rarefied electron density, called a *Fermi hole*. Each hole is equivalent to a positive charge accompanying each electron and this contributes a new term (negative term) to the potential energy.

5.4 Orbital energies

The Hartree–Fock SCF orbitals φ_i considered so far have an energy associated with them. This means that they diagonalize the Hartree–Fock hamiltonian matrix (cf. page 21):

$$\langle \varphi_i | \mathscr{H}_{HF} | \varphi_j \rangle. \tag{5.31}$$

However, as already found, a unitary transformation of a set of spin-orbitals leaves the wavefunction and the total energy unchanged. This means that there is not a unique set of monoelectronic SCF functions. But only those that satisfy

$$\langle \varphi_i | \mathscr{H}_{HF} | \varphi_i \rangle = \varepsilon_i \qquad \langle \varphi_i | \mathscr{H}_{HF} | \varphi_i \rangle = 0 \tag{5.32}$$

can have an energy associated with them. They are called *canonical orbitals*.

As mentioned at the beginning of this chapter, the mono-ionization potentials – obtained for example by photoelectron spectroscopy – provide experimental evidence for the number of occupied energy levels and for the relative positions of these levels (the expression energy level is used here in a somewhat loose way, see Section 5.6). They also lead to values for the energy E_i of each bound electron (although only approximately, as will be seen below).

The ionization potential I_i for a given bound electron is the minimum energy needed to remove the electron from the atom or molecule (of a gaseous sample), that is the energy that must be given to the electron so that both its kinetic energy and its potential energy in the field of the nucleus become zero:

$$X \rightarrow X^+(-i) + e_i^- \qquad \Delta U_i = -I_i. \tag{5.33}$$

If that was the only change, then

$$E_i + I_i = 0 \qquad E_i = -I_i. \tag{5.34}$$

However, there are simultaneous changes in the energies of the remaining electrons. Therefore, expression (5.34) is only approximate. In addition, it must be recognized that the positive ion $X^+(-i)$ may appear in different states (see Section 5.6) and that, if the i-th electron is an inner electron, an orbital transition may occur. When the removed electron is strongly bound, such as the core electrons of atoms of high nuclear charge, an important source of error comes from relativistic effects (see Section 5.8).

Within the SCF HF framework, the energy ε_i, as a diagonal element of the Hartree–Fock matrix (5.31), can be shown to be equal to the difference between the expectation values of the hamiltonian for the neutral species X (the Hartree–Fock total energy) and for a positive ion $X^+(-i)$ described by a Slater determinant built on the basis of canonical orbitals identical to those of X (frozen orbital model). This apparently crude description of the $X^+(-i)$ can be shown to give the lowest energy for the ion. As a result, ε_i is an approximate estimate of the ionization potential of the i-th electron:

$$\varepsilon_i \cong -I_i. \tag{5.35}$$

This conclusion was originally pointed out and discussed by the Dutch physicist T.A. Koopmans in 1933. It is known as *Koopmans theorem* (for a recent criticism of the current didactic presentation, see ref. 46). The result (5.35) can, alternatively, be attributed to compensation of the two possible sources of error: neglect of reorganization of the remaining electronic charge through orbital relaxation (stabilization of the positive ion) and neglect of the change (increase) in electron correlation energy on going from the neutral to the positively charged species (see, for example, ref. 13).

We have already stated that 2s and 2p orbitals have different energies because the repulsion suffered by a 2s electron is different from that of a 2p electron. The reason why it is smaller for 2s lies partly in the following qualitative considerations due to the American chemist John Slater (1900–1976) (ref. 47).

Any electron i at a given distance r from a nucleus, besides feeling the nuclear attractive force, experiences repulsion from all the electrons that lie inside and outside a spherical surface of radius r. Any of these repulsions affects the energy of electron i. However, as far as the electric field at the position of electron i is concerned, there is a difference arising from electrostatics, as follows:

(a) the net effect of a spherical electron distribution outside the sphere of radius r vanishes.
(b) the net effect of the electrons inside the sphere is equivalent to the effect they would have if they were concentrated at the centre (the nucleus).

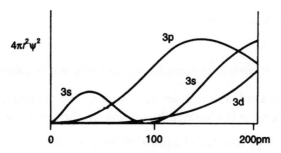

Fig. 5.4 Radial probability distributions of hydrogen-like 3s, 3p and 3d orbitals for atom K.

The value of an ns orbital close to the nucleus is higher than that of an np orbital of the same atom; s orbitals are *more penetrating* than p orbitals of equal quantum number n (recall that p orbitals have a node at the position of the nucleus). As a consequence, there is more negative charge to be taken into account for effect (b) in the case of a p orbital. The same is true of d orbitals with respect to p orbitals of equal n. Figure 5.4 shows the radial probability distributions of the degenerate hydrogen-like 3s, 3p and 3d orbitals for the potassium atom.

Therefore, *if* the contributions from the outer electrons to the energy of s, p or d electrons are not very much different – as is expected to be the case if the s, p, d electrons are valence-shell electrons – then the order of penetration $ns > np > nd$ implies an opposite order for the energies $ns < np < nd$. However, the problem is not as simple as it seems. In particular it should be stressed that the averaged potential energy associated with a given electrostatic interaction depends on the mean of (1/distance) and not on (1/mean distance). In addition, changes in kinetic energy must also be considered (for a recent discussion, see ref. 161 and letters that followed in the 1999 May issue of the *Journal of Chemical Education*).

In 1930, Slater proposed an approximate quantitative way of dealing with these effects: to replace the nuclear attraction and the repulsions from the inner electrons by a mitigating attraction to a pseudo-nucleus of charge

$$Z^*e = (Z - \sigma)e. \tag{5.36}$$

The σ parameter is called the *screening constant* and Z^*e is the *effective nuclear charge*. Of course, the screening effect by the inner electrons is not to be regarded as a way of reducing the actual attractive effect of the nucleus but as an additional effect of opposite sign.

Problem 5.4   For which of the orbitals 3s, 3p, 3d is the screening constant largest? Why?

Problem 5.5   Is it correct to consider σ_{2p} as being independent of Z? Justify.

Slater put forward a few rules to estimate the screening constant. For *n*s and *n*p electrons, σ is obtained by adding up the following contributions:

(a) 0.35 for any other electron of equal n (0.30 for $n = 1$);
(b) 0.85 for each electron of principal quantum number $n - 1$;
(c) 1.00 for each electron of principal quantum number $n - 2, n - 3, \ldots$

Problem 5.6
(a) *Calculate Z^* for electrons 2p and 3s of Na in its lowest energy configuration.*
(b) *Find that, for valence electrons, Z^* increases along the second period of the Periodic Table, but decreases on going from $Z = 10$ (Ne) to $Z = 11$ (Na).*

The Slater Z^* values (effective atomic number) are useful in the interpretation of differences between atomic radii and between electronegativities. They can also be used in qualitative discussions of ionization potentials (see Problem 5.7). For example, the overall trend of an increase of ionization energy across a period of the Periodic Table is attributed to the fact that each successive electron which is added ineffectively shields the other electrons from the increasing nuclear charge. However, they are less satisfactory in other respects. For example, according to the rules above, Z^* is the same for *n*s and *n*p electrons and yet the s orbital is more penetrating than p. It is only for *n*d that the extended rules forsee a larger screening constant than for *n*s and *n*p. But the major errors occur when the Slater rules are taken in absolute terms, for example in the estimation of ionization energies by using a modified Bohr equation

$$I = \mathcal{R} Z^{*2} / n^2 \qquad\qquad (5.37)$$

$\mathcal{R}$ (Rydberg constant) $= 1312\,\text{kJ mol}^{-1}$.

For the first ionization energy of oxygen, the calculated value is $6780\,\text{kJ mol}^{-1}$ and the experimental value is $1314\,\text{kJ mol}^{-1}$.

Problem 5.7   Compare the quotient of the second and first ionization energies of O, calculated with Eqs. (5.36) and (5.37), with the experimental value 2.6.

An alternative characterization of screening constants based on the experimental ionization energies themselves can be found in ref. 48.

The concept of effective nuclear charge serves as the basis for the definition of functions in analytical form which approximately reproduce the numerical SCF HF orbitals. They are called *Slater orbitals*. The radial parts of these orbitals have the general form

$$R(\text{SO}) = Nr^{n-1}e^{-\xi r}$$
$$\xi = Z^*/n \tag{5.38}$$

whereas the angular part is identical to that for monoelectronic atoms, $Y(\theta, \phi)$. Slater-type orbitals (STO) are frequently used as a starting point in SCF HF calculations both in atoms and molecules (see Chapter 7).

5.5 Electronic configurations

On the basis of the well-known orbital energy sequence within the SCF HF approximation

$$1s < 2s < 2p < 3s < 3p < 3d, 4s < \cdots \tag{5.39}$$

the electronic configurations of atoms of small atomic number can be directly established by filling the various orbitals successively, obeying the Pauli principle of two electrons for each orbital: the 'building up' ('aufbau', in German) method. For example,

$$_{11}\text{Na}: 1s^2, 2s^2, 2p^6, 3s^1. \tag{5.40}$$

Problem 5.8 Interpret the following plots for the successive ionization energies (I) of atoms of low atomic number, taken from refs. 49 (with permission from Education in Chemistry, *1990) and 50 (with permission from* Journal of Chemical Education, *1991), respectively.*

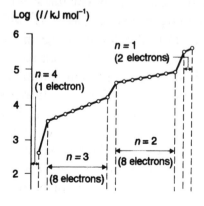

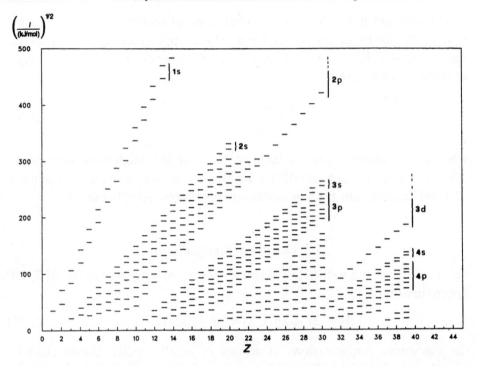

Implicit in this method is the (wrong) assumption that the total energy is the sum of the individual energies. Expression (5.26) states where the difference lies. However, because of the comparatively large energy difference for orbitals of low principal quantum number, the building up method works for small atomic numbers, but problems may arise especially when dealing with transition metal atoms for which valence nd and $(n+1)$s orbitals become close in energy.

Figure 5.5 schematically shows the energies of various orbitals as a function of atomic number.

It should be noted that $\varepsilon_{3d} < \varepsilon_{4s}$ except for $Z=19$ and $Z=20$. Accordingly, the valence sub-configuration for $_{19}$K is 4s^1 and not 3d^1, and that for $_{20}$Ca is 4s^2 and not 3d^2. For $_{21}$Sc, the building up method would predict a valence sub-configuration 3d^3 but experiment requires 3d^1, 4s^2. Another example: the predicted 3d^8 valence sub-configuration of $_{26}$Fe is at variance with the experimental data and with full theoretical calculations which point to 3d^6, 4s^2 (for recent discussions, see refs. 51 and 52). That is, 'higher orbitals are occupied despite vacancies in lower ones' (ref. 53).

One should be prepared for these discrepancies on account of Eq. (5.26), but some physical insight can be gained from the following discussion (refs. 53 and 54). First of all, it cannot be forgotten that the energy

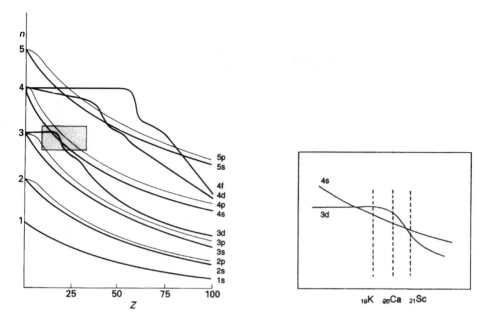

Fig. 5.5 Schematic representation of the relative orbital energies as a function of atomic number.

associated with each orbital depends on the configuration, that is, it depends on the occupation number of the orbitals including itself. We recall that the Schrödinger equation defines the energy levels of a one-electron atom without regard to which level may be occupied, whereas the SCF equations define orbital energies in a many-electron atom only with reference to a particular set of occupation numbers. For example, for $3d^1$, $4s^2$, ε_{3d} is the energy of an electron in the field created by the atomic core and two 4s electrons; it is different from ε_{3d} for the configuration $3d^2$, $4s^1$ or $3d^3$. The same can be said about ε_{4s}. Therefore, in this case (scandium) there are five orbital energy values to be considered and not just two.

Since a 4s orbital is more diffuse than a 3d orbital, 3d,3d repulsions have a larger positive contribution to a SCF HF energy than do 4s,4s or 3d,4s repulsions. For an atom or ion of a transition element, the repulsion sequence is

$$4s,4s < 3d,4s < 3d,3d. \qquad (5.41)$$

As a consequence, the energies ε_{3d} and ε_{4s} both increase with the 3d population. For example, a 'transition' from $3d^n$, $4s^2$ to $3d^{n+1}$, $4s^1$ is accompanied by the substitution of a 4s,4s repulsion by a 3d,4s repulsion which increases both ε_{3d} and ε_{4s}. Figure 5.6 shows this effect for $_{21}$Sc.

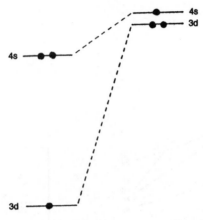

Fig. 5.6 Orbital energy changes for the valence configurations $3d^1$, $4s^2$ and $3d^2$, $4s^1$ of $_{21}$Sc (ref. 54).

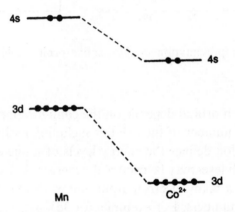

Fig. 5.7 Orbital energy changes for the valence configuration $3d^5$, $4s^2$ of $_{25}$Mn and $_{27}$Co^{2+}: the effect of a higher effective nuclear charge (ref. 54).

Problem 5.9   Using the Slater rules, calculate Z^* for the 4s orbital for a valence configuration $3d^p$, $4s^q$, and conclude that a $4s \rightarrow 3d$ 'transition' leads to a decrease of Z^* for both 4s and 3d.

In apparent contradiction of the above discussion, the metal (positive) ions do have configurations according to the 'aufbau' procedure. For example, for Fe^{2+} it is $3d^6$ and not $3d^4$, $4s^2$. The explanation is found in the increase of Z^* which leads to a stabilization of the 3d and 4s orbitals and to an increase in their energy separation. Figure 5.7 depicts the energies for 3d and 4s in the cases of $_{25}$Mn and $_{27}$Co^{2+} which have a common configuration: $3d^5$, $4s^2$.

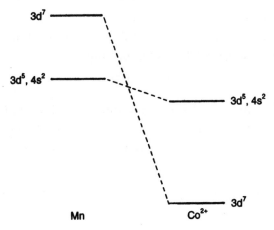

Fig. 5.8 Comparison between the energies for the configurations $3d^5$, $4s^2$ and $3d^7$ of $_{25}$Mn and $_{27}$Co^{2+} (ref. 54).

In Fig. 5.8, a comparison is made between the energies for the configurations $3d^5$, $4s^2$ and $3d^7$ of $_{25}$Mn and $_{27}$Co^{2+}.

The above discussion is not complete. There is the additional contribution of the changes in exchange energies from one configuration to another (ref. 51), as well as the changes in the energy of the core (ref. 52). Also, configuration energies considered so far are average values for the various energy levels associated with each open-shell configuration, a matter we shall deal with in the next section.

5.6 Beyond electronic configurations: terms, levels, states

The spin–orbit interaction considered in the study of monoelectronic atoms also plays an important role in polyelectronic atoms, with the exception of closed-shell configurations for which the total orbital angular momentum and the total spin angular momentum vanish.

As the core electrons are a closed sub-shell, the magnetic moments associated with their orbital and spin angular momenta cancel out, and therefore only the valence electrons need be considered. For atoms with just one valence electron, the situation is similar to that for monoelectronic atoms. For example, for the excited configuration of $_{19}$K, [Ar], $4p^1$, we have

$$\ell = 1$$
$$s = 1/2 \tag{5.42}$$
$$j = 1 + 1/2 = 3/2 \text{ and } 1 - 1/2 = 1/2.$$

There are two possibilities for the spin–orbit interactions, corresponding to two values of the total angular momentum quantum number j. Therefore, two energy levels exist for such a configuration:

$$4^2P_{1/2} \quad 4^2P_{3/2}. \tag{5.43}$$

The superscript on the left of P is the number of values m_s ($2s+1=2$) (spin multiplicity) and the remaining indices have the same meaning as on page 69. The spin–orbit splitting is larger for K than for Na: it is roughly dependent on Z^4.

Problem 5.10   *Calculate the difference in energy between the levels (5.43) using the expression*

$$E_{l,s,j} = hcA[j(j+1) - \ell(\ell+1) - s(s+1)]/2$$

where the spin–orbit coupling constant A is $38.5\,\mathrm{cm}^{-1}$.

Problem 5.11   *Interpret the diagram below regarding the D lines of the sodium emission spectrum.*

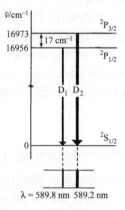

The spin–orbit interactions are weak for light atoms but can overcome the electrostatic interactions for atoms of high atomic number. The spin–orbit coupling is only one of the factors determining the occurrence of several energies for a given open-shell configuration. The other is electron repulsion. Let us consider the carbon configuration as an example:

$$1s^2, 2s^2, 2p^2. \tag{5.44}$$

Depending on the spin states of the two p electrons and the way they are distributed among the three 2p orbitals (six spin-orbitals), 15 possibilities (combinations) exist. There are, then, 15 states. Here are some of the combinations:

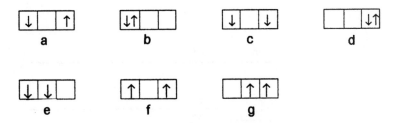

Obviously, states (a) and (b) differ in the electron repulsion energy, hence have different energies. We say that they belong to different *terms*. On the other hand, states (a) and (c), although belonging to the same term, differ in energy because of a different spin–orbit coupling. We say that they belong to different *levels*.

The atomic terms are characterized by the orbital angular momentum which is reflected in the value of the total orbital quantum number L. Just as the individual angular momenta are quantized, so are the resultants. For the example above, with $\ell_1 = \ell_2 = 1$ for two p electrons, the maximum value of the resultant angular momentum is represented by the sum of the quantum numbers, and the minimum value is represented by the difference. Intermediate values are allowed too, provided they differ from the extremes by a whole number:

$$L = \ell_1 + \ell_2, \ell_1 + \ell_2 - 1, \ldots, \ell_1 - \ell_2. \tag{5.45}$$

Thus,

$$L = 2, 1, 0$$

which corresponds to terms designated, respectively, by

$$D, P, S$$

(by analogy with d, p, s orbitals for $\ell = 2, 1, 0$).

The atomic levels of each term are defined according to the total angular momentum which is reflected in the value of the quantum number J. This can be expressed as a function of L and S

$$J = L + S, \; L + S - 1, \ldots, L - S \tag{5.46}$$

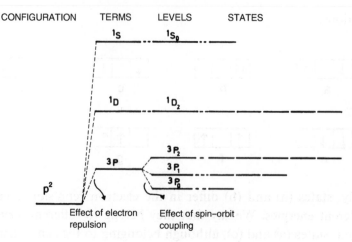

Fig. 5.9 Terms, levels and states for a p^2 configuration.

with

$$S = s_1 + s_2, s_1 + s_2 - 1, \ldots, s_1 - s_2$$

$$\quad 1 \qquad\qquad 0 \qquad\qquad 0 \tag{5.47}$$

Figure 5.9 presents the several terms and levels for the $2p^2$ sub-configuration allowed by the Pauli principle. The total spin multiplicity $2S+1$ is also shown: for $S=0$, the *singlet states* ^{1}S and ^{1}D; for $S=1$, the *triplet states* ^{3}P.

Problem 5.12   Find all the levels for the configuration $2p^1$, $3p^1$, for which there are no restrictions imposed by the Pauli principle.

The energy order for the different terms and levels in Fig. 5.9 is in accord with the so-called *Hund's rules* put forward by the German spectroscopist Friedrich Hund (1896–1998), as follows:

(a) the term with the highest spin multiplicity (the highest total spin S) lies lowest in energy.
(b) for the same spin multiplicity, the term with the highest quantum number L lies lowest in energy.
(c) for a given term, the levels with lower quantum number J have lower energy if the valence shell is less than half full.

Belonging to the triplet P term, ^{3}P, there are states such as

where the two electrons have parallel spins. These are states of lower energy than, for example,

However, it should be noted that, in term ^3P, states are included for which we cannot speak of parallel spins. Figure 5.10 shows the electron distributions for the several states belonging to the term ^3P (in some cases, a combination of distributions).

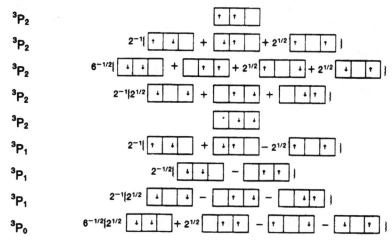

Fig. 5.10 States for the term ^3P of configuration p^2 (reprinted with permission from *Journal of Chemical Education*, **68**, 134 (1991); copyright 1991, Division of Chemical Education Inc., ref. 55).

The origin of rule (a) lies in the electron spin correlation associated with the antisymmetry of the wavefunction. In particular, the difference in energy between a singlet state and a triplet state comes mainly from the exchange energy (see the example of He below). Since electrons of the same spin tend to keep apart, it is usually considered that their Coulomb repulsion is henceforth decreased. However, some rigorous calculations (refs. 16 and 56) indicate that the situation may not be as simple as that, which in fact should not be surprising because electrostatic repulsion varies with $1/r_{ij}$ and the mean value of $1/r_{ij}$ is not the inverse of the mean value of r_{ij}. Moreover, detailed calculations for He have shown that the repulsion between the electrons is greater in triplet than in singlet states (ref. 57 and references therein). The lowering of the energy in the former is due to orbital contraction, giving an increase in the electron–nucleus attraction.

Regarding rule (b), a physical explanation has been put forward (see, for example, ref. 16): electrons circulating in opposite directions, with a low total orbital angular momentum, will meet frequently, and so have a large repulsive interaction.

The explanation of rule (c) is related to spin–orbit coupling. For an electron in an atom, the energy is lowest when the spin and the orbital magnetic moments are opposed. This implies a low total angular momentum. The inversion of the levels when the shell is more than half full reflects a change of sign of the spin–orbit coupling constant.

Problem 5.13   *Identifying the lowest energy term of $_7N$ as ^4S, explain why the ionization energy of nitrogen is larger than those of carbon and oxygen.*

Let us now consider, in some detail, the singlet and the triplet states of He for the excited configuration $1s^1$, $2s^1$. There are four states, depending on the spins of the electrons. The antisymmetry of the wavefunctions can be accomplished either through the spatial part or the spin part, respectively:

$$[1s(1)2s(2) - 1s(2)2s(1)]/\sqrt{2} \tag{5.48}$$

$$[\alpha(1)\beta(2) - \alpha(2)\beta(1)]/\sqrt{2}. \tag{5.49}$$

Thus, the singlet state is represented by

$$\psi_s = [1s(1)2s(2) + 2s(1)1s(2)][\alpha(1)\beta(2) - \beta(1)\alpha(2)]/2 \tag{5.50}$$

whereas the three functions for the triplet state (a triplet of states) are

$$\psi_T = [1s(1)2s(2) - 2s(1)1s(2)]/\sqrt{2} \begin{cases} [\alpha(1)\alpha(2)] \\ [\alpha(1)\beta(2) + \beta(1)\alpha(2)]/\sqrt{2} \\ [\beta(1)\beta(2)]. \end{cases} \tag{5.51}$$

Note that the second function ψ_T does not, rigorously, entitle us to speak of parallel spins. Such a function requires two Slater determinants:

$$(1/2) \begin{vmatrix} 1s(1) & 1s(2) \\ 2\bar{s}(1) & 2\bar{s}(2) \end{vmatrix} + (1/2) \begin{vmatrix} 1\bar{s}(1) & 1\bar{s}(2) \\ 2s(1) & 2s(2) \end{vmatrix} \tag{5.52}$$

Problem 5.14
(a) *Confirm the statement that a triplet state does not necessarily mean two parallel spins.*
(b) *Find for Eq. (5.50) an expression similar to Eq. (5.52).*

The energies of the two states for this open-shell configuration are

$$E_S = \varepsilon_{1s}^0 + \varepsilon_{2s}^0 + J + K$$
$$E_T = \varepsilon_{1s}^0 + \varepsilon_{2s}^0 + J - K \tag{5.53}$$

where ε_{1s}^0 and ε_{2s}^0 are the energies of the 1s and 2s orbitals of He^+. The Coulomb integral, J, and the exchange integral, K, have a form equivalent to Eqs. (5.27) and (5.28), respectively. Since J and K are positive quantities, $E_T < E_S$.

In the examples given, the effect of spin–orbit coupling is less than that of electron repulsion. This allows the individual orbital angular momenta l_i and the individual spin angular momenta s_i to be added separately to obtain the respective resultant vectors L and S, which are added together to yield the total angular momentum J. This is called the *Russell–Saunders* or *L–S coupling* (for a recent analysis of L–S coupling see ref. 58.)

For atoms of high atomic number, spin–orbit coupling becomes more important. As a consequence, each l_i and each s_i are first added to yield j_i, which are then added leading to J (*jj coupling*). For many atoms an intermediate situation is encountered (see, for example, ref. 59).

The low-energy terms for all the atoms up to $Z = 103$ and their periodicity have been studied in ref. 60.

5.7 Density-functional theory and Kohn–Sham orbitals

Density-functional theory goes beyond the Hartree–Fock approach in that correlations are taken into account which are not contained in the SCF HF energy. The basic idea of density-functional theory (DFT) is almost as old as quantum mechanics: to relate directly the total energy of an atomic or molecular system to the ground state electron density $\rho(\mathbf{r})$ as given in diffraction experiments, instead of involving the system wavefunction. Mathematically speaking, the energy is assumed to be a *functional* of $\rho(\mathbf{r})$, $E[\rho]$, which involves an integral over an expression depending only on $\rho(\mathbf{r})$:

$$E_{\text{tot}} = \int \mathcal{E}[\rho(\mathbf{r})] \, d\mathbf{r} = E[\rho]. \tag{5.54}$$

In 1964, Pierre Hohenberg and Walter Kohn proved that the exact ground state energy of a correlated electron system is a functional of the density $\rho(\mathbf{r})$ and that this functional has its variational minimum when evaluated for the exact ground state density. This means that a variational equation would lead to a knowledge of the exact ground state energy and density (and hence any ground state property), if the exact functional $E[\rho]$ was known. As this is not the case, certain approximations are needed.

The idea that made DFT calculations actually feasible for atoms (as well as molecules) is that an electron density can be represented as a sum of individual orbital contributions. This is the Kohn–Sham method which establishes a formal equivalence between the determination of the ground state energy, including many-body effects, and the Hartree-like problem of the self-consistent solution of a set of single-particle equations. As a result, a set of occupied orbitals are obtained as auxiliary functions that add up to yield the total electron density. Given the auxiliary nature of the *Kohn–Sham orbitals*, one should expect no simple physical meaning for the orbital energies. In fact, only the (exact) energy of the highest occupied orbital can be equated to the symmetric value of an ionization potential (the lowest one), whereas minus the energy of the lowest unoccupied orbital is the electron affinity. This method has also been applied to the study of many types of chemical problems, from calculating the geometrical structures of molecules to explaining the pathways of chemical and enzymatic reactions (for a recent review, see ref. 61). This can be accomplished at a fraction of the computing cost of Hartree–Fock calculations. In addition, density-functional models provide very good descriptions of systems which are not adequately described by Hartree–Fock methods, such as transition metal and organometallic compounds. Meanwhile, user-friendly software products have been made available for DFT calculations using common microcomputers. The importance of density-functional theory to chemistry was recently acknowledged when the American chemist Walter Kohn (1923–) shared the 1998 Nobel Prize in chemistry for his contribution to the development of the theory.

5.8 Relativistic corrections

Spin–orbit coupling is an addition to the Schrödinger equation but it is a natural feature in Dirac's theory which associates relativity theory with quantum mechanics. There are, however, other relativistic effects in the electronic structure of polyelectronic atoms which can be related to changes in the electron mass with velocity (for a review on relativistic effects in structural chemistry, see ref. 62).

Near the nucleus, the electrons – especially the s electrons which have a maximum probability density for points near the nucleus – accelerate and their relativistic mass increases. The result is a decrease of the average distance to the nucleus (see the expressions (3.29) and (3.30) and page 53 for the expression of the Bohr radius as the radius of the first Bohr orbit for H). That contraction means a decrease of energy.

This effect is more pronounced for atoms of high nuclear charge, where the kinetic energy of the s electrons is very high close to the nucleus. Because the higher s orbitals have to be orthogonal against the lower ones, they will suffer an additional contraction; their contraction can in fact be even larger. It is this effect that explains the smaller atomic size of the alkaline metal francium, $_{87}$Fr, comparing with caesium, $_{55}$Cs, so opposing the general increase of atomic radius down a group of the Periodic Table. In the same way, the inversion of the first ionization energies in group 11 can be interpreted: $I_{Cu} > I_{Ag} < I_{Au}$; in any case it is a valence s electron that is being removed, but the 6s valence electron of Au is especially stabilized due to the increase of its relativistic mass.

The stabilization of 6s relative to the 5d orbitals of Au helps to explain the yellow colour of gold thanks to the absorption of blue light for excitation of electrons from the 5d band to the 6s band (see Chapter 11),

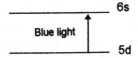

whereas silver only absorbs ultraviolet radiation (ref. 63 and references therein):

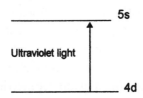

The fact that the dissociation energy of Au$_2$(g) is larger than that of Ag$_2$(g) can also be interpreted on the basis of 6s orbital stabilization of Au. For the same reason, the diatomic ion Hg$_2^{2+}$, which is isoelectronic of Au$_2$, is stable relative to Hg$^+$.

Problem 5.15   Interpret the occurrence of mercury as a liquid metal on the basis of the small participation of the 6s orbitals in the metallic bonding (see, also, ref. 64).

6

Orbitals in diatomic molecules

6.1 The approximations

When extending the molecular orbital concept developed for the monoelectronic species H_2^+ to polyelectronic diatomic molecules, we start by acknowledging the role of two fundamental approximations: (a) one associated with the existence of two nuclei as attractive centres, namely the *Born–Oppenheimer approximation*, as already encountered in H_2^+; and (b) the other related to the concept of the orbital when two or more electrons are present, that is the neglect of the *electron coulomb correlation*, as already discussed on going from mono- to polyelectronic atoms. Within the orbital approach, an additional feature when comparing to H_2^+ is the exchange energy directly associated with the Pauli principle.

The Born–Oppenheimer approximation treats the nuclear motion and the electron motion as entirely independent, which is reasonable in view of the huge difference of mass for nuclei and electrons. This means that the electronic energy is calculated for a given geometric arrangement of the nuclei and the internuclear repulsion is then added as a separate term.

To replace the real repulsions between electrons, for every instantaneous electron–electron distance, by average values, although generated by the best self-consistent field calculation, ignores the fact that the positions of the electrons for each pair are electrostatically correlated and this affects the probability distribution of each electron and the total electronic energy. The correlation energy error results from electron pair effects and, thus, is reasonably constant for geometric changes that leave the electron pairings essentially unaltered, such as in conformational changes. Errors can, however, be quite serious in calculations of ionization potentials and of dissociation energies. One way of reducing the correlation errors, within the framework of the orbital concept, is to mix the first wavefunction found with wavefunctions appropriate for excited states of the same symmetry: *configuration*

114

interaction. Another, increasingly used way, is *density-functional theory* (see Section 5.7).

Just as for polyelectronic atoms, the electronic wavefunction for a molecule must be antisymmetric: the Pauli principle. Thus, *electron spin correlation* is accommodated by the definition of the wavefunction as a Slater determinant whose elements are the occupied molecular orbitals.

Relativistic corrections are usually neglected and this is acceptable whenever we have elements of not too high an atomic number and providing that we are dealing with valence electrons which have a small probability of being close to the nuclei, as is often the case.

Since an analytical solution of the Schrödinger equation is no longer possible, the molecular orbitals must be constructed as linear combinations of basis functions using the variation principle. An exact solution would require an infinite set of basis functions, leading to an infinite set of m.o.s. Thus an additional approximation, at the operational level, is introduced which is dependent on the dimension of that set. Further approximations are encountered in the calculation of the m.o.s, depending on the estimation of the various H_{ij} integrals as matrix elements as required by the variation theorem. We shall develop this subject at the end of Chapter 7.

In the present chapter, we will explore the potential of molecular orbital theory at an essentially qualitative level, based on symmetry considerations for homonuclear diatomic molecules. By taking advantage of these symmetry relations, molecular orbitals are built as linear combinations of the valence atomic orbitals of the intervening atoms. A general rule of conservation is observed: *the number of m.o.s is always equal to the number of a.o.s we start with.* Both the essential shape of m.o.s and their relative energies can be obtained in this simplified but analytical manner. As a consequence, electronic configurations can be established that enable relative values of bond energies and bond lengths to be understood, as well as magnetic properties. A direct extension to polar bonds is also possible.

6.2 The simple diatomics H_2, $He_2{}^+$ and 'He_2'

By comparing the bond energies in $H_2{}^+$ and H_2 and the respective bond lengths

	$H_2{}^+$	H_2
Bond energy	$255\,\text{kJ mol}^{-1}$	$432\,\text{kJ mol}^{-1}$
Bond length	$106\,\text{pm}$	$74\,\text{pm}$

it can be concluded that each electron of ground state H_2 has a net bonding effect. It is also found that there is only one ionization energy for ground state

H_2 and that the molecule is diamagnetic, that is the electron magnetic moments cancel each other out. This means that the two electrons occupy the same (bonding) molecular orbital ψ_1, with opposite spins. Therefore, the ground state electronic wavefunction Ψ is given by the following Slater determinant:

$$\Psi = (1/\sqrt{2}) \begin{vmatrix} \psi_1(1) & \psi_1(2) \\ \bar{\psi}_1(1) & \bar{\psi}_1(2) \end{vmatrix}. \tag{6.1}$$

An approximate form for ψ_1 is identical to Eq. (4.7) for H_2^+ in terms of the linear combination of the 1s orbitals of isolated H atoms, A and B, that is a σ molecular orbital:

$$\sigma_1 = (\varphi_{1sA} + \varphi_{1sB})/\sqrt{2(1+S)}. \tag{6.2}$$

The corresponding probability density is

$$\sigma_1^2 = (\varphi_{1sA}^2 + \varphi_{1sB}^2 + 2\varphi_{1sA}\varphi_{1sB})/[2(1+S)]. \tag{6.3}$$

S is the overlap integral for the atomic orbitals φ_{1sA} and φ_{1sB}, at the internuclear distance of 74 pm. As we have seen in Section 4.3 for H_2^+, the interpretation of bond stabilization requires consideration of changes in both the electronic potential and kinetic energy, and atomic orbital contraction plays an important role (ref. 39). This role is simulated by assuming a larger effective nuclear charge in the mathematical expressions of φ_{1sA} and φ_{1sB} (see Table 3.1 and expression 5.38).

The electron probability density along the line passing through the nuclei is graphically represented as in Fig. 4.5 for H_2^+ and the isoprobability contours for a plane containing the internuclear axis are similar to those of Fig. 4.7 for the same ion. If we seek a distribution similar to the radial probability distribution for atoms, Fig. 6.1 is obtained (ref. 65). It shows the circular distribution of electron density for different distances from the internuclear axis. It is found that the electronic charge is concentrated in a circular 'doughnut' around the H–H axis, with a maximum at about 37 pm from the axis and about 50–55 pm from each nucleus.

The corresponding anti-bonding m.o. is

$$\sigma_1' = (\varphi_{1sA} - \varphi_{1sB})/\sqrt{2(1-S)} \tag{6.4}$$

and the corresponding circular distribution is shown in Fig. 6.2.

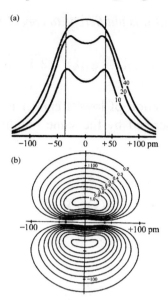

Fig. 6.1 Circular distribution of the electronic probability for the bonding m.o. of H_2 for increasing distances from the internuclear axis, and corresponding contours in a plane containing that axis (adapted with permission from ref. 65, *Journal of Chemical Education*, **68**, 743 (1991); copyright 1991, Division of Chemical Education Inc.).

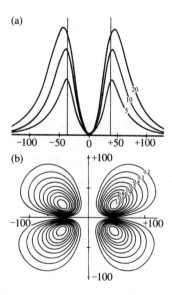

Fig. 6.2 Circular distribution of the electronic probability for the anti-bonding m.o. of H_2 for increasing distances from the internuclear axis, and corresponding contours in a plane containing that axis (adapted with permission from ref. 65, see Fig. 6.1).

Problem 6.1   Confirm that it is indifferent to consider Eq. (6.4) or

$$\sigma'_1 = (\varphi_{1sB} - \varphi_{1sA})/\sqrt{2(1-S)}.$$

The electronic configuration of lowest energy is $(\sigma_1)^2$, the electrons with opposite spins being described by the same bonding m.o. σ_1. Graphically, we have

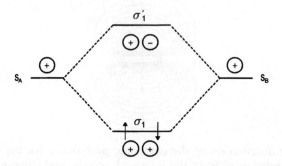

or more rigorously

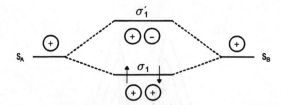

when due account is taken of the normalization factors (see page 79). The fact that the ionization energy of H_2 (1491 kJ mol^{-1}) is larger than that for atomic hydrogen (1312 kJ mol^{-1}) is in agreement with the diagram above. For a recent computer program which displays potential energy curves, molecular orbitals and electron densities, see ref. 37.

The result for H_2 is immediately extended to the diatomic species He_2^+, for which the approximate m.o.s are again constructed from two 1s atomic orbitals, the (modified) valence a.o.s of He atoms. The bond energy is 251 kJ mol^{-1}, which is smaller than for H_2 and close to the value for H_2^+. In agreement with this, the bond length, 108 pm, is larger than the value for H_2 and near the value for H_2^+. Part of these changes must be due to a greater internuclear repulsion as a consequence of higher nuclear charges. The other part is the result of the anti-bonding character of one of the three electrons:

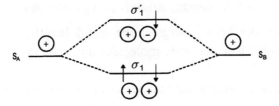

For a hypothetical molecule 'He$_2$', the electron assignment would be

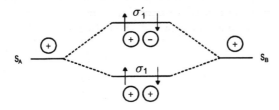

which shows why there are no (ground state) diatomic helium molecules. The fact that the anti-bonding m.o. is more anti-bonding than the bonding m.o. is bonding justifies the non-association of two He atoms: the energy of 'He$_2$' is larger than the energy of two He atoms in their ground states. This effect is sometimes called a Pauli repulsion, in so far as the filling of the anti-bonding m.o. is a consequence of the application of the Pauli principle to the bonding m.o. However, it should be noted that, for internuclear distances of about three times the typical bond length values, the energy of two He atoms is slightly smaller than for infinite distance, due to London dispersion interactions. A transient van der Waals diatomic species does exist. Figure 6.3 compares the change of energy of two H atoms as a function of distance with the corresponding variation for two He atoms.

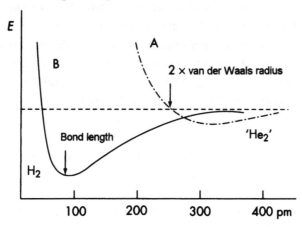

Fig. 6.3 Energy for two He atoms (A) and for two H atoms (B) as a function of the internuclear distance.

6.3 Molecular orbitals in X_2 molecules

Table 6.1 shows the bond energies, the bond lengths and the magnetic features of homonuclear diatomic molecules and ions involving second row elements.

We will see how a qualitative m.o. description can explain the differences found in Table 6.1.

Table 6.1 *Characteristics of homonuclear diatomic species*

Molecule	Bond energy/kJ mol^{-1}	Bond length/pm	Magnetism
Li_2	105	267	Diamagnetic
Be_2	55	?	Diamagnetic
B_2	293	159	Paramagnetic
C_2	602	124	Diamagnetic
N_2^+	841	112	Paramagnetic
N_2	942	110	Diamagnetic
O_2^+	646	112	Paramagnetic
O_2	494	121	Paramagnetic
O_2^-	393	132	Paramagnetic
O_2^{2-}	150	149	Diamagnetic
F_2	155	141	Diamagnetic

The involvement of second row atoms brings a new problem: more than one atomic orbital contributed by each atom. The normal procedure is to regard the core orbital (1s atomic orbital) as unaltered by the bonding of the two atoms, hence to construct the molecular orbitals on the basis of the valence atomic orbitals only. This approximation is justified on the basis of the much lower energy of the 1s electrons with respect to the valence electrons and their more compact arrangement close to the nucleus. This approximation is acceptable for many purposes. A brief test of its general validity is to compare the maximum values for the mono-ionizations of atom X (removal of a 1s electron) and molecule X_2. For example, for N and N_2 the values are, respectively, 38.8×10^3 kJ mol^{-1} and 39.5×10^3 kJ mol^{-1}. The difference between them is relatively small (about 2%). However, differences like this play a central role in X-ray photoelectron spectroscopy (XPS or electron spectroscopy for chemical analysis, ESCA) which is based on the removal of the least energetic core electrons. Figure 6.4 illustrates this effect with the case of N_3^- (see, for example, ref. 66).

The molecule Li_2

This example will be discussed in some detail as it will serve as a paradigm for the cases to follow. It will largely resemble the discussion of excited

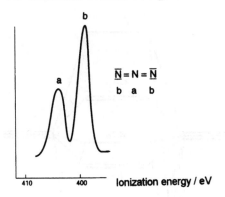

Fig. 6.4 Photoelectron spectrum for the inner electrons of N_3^-, corresponding to the slightly perturbed 1s electrons of the nitrogen atoms.

m.o.s of H_2^+ carried out in Chapter 4. The dilithium molecule exists in lithium vapour and its bond can be described by assigning its two valence electrons (note that Li has the ground state configuration $1s^2$, $2s^1$) to a bonding m.o. Since the two electrons must then have antiparallel spins, the diamagnetism exhibited by this species is thus explained. To a first approximation, the bonding m.o. could be constructed from the 2s atomic orbitals:

$$\sigma_1 = (\varphi_{2sA} + \varphi_{2sB})/\sqrt{2(1+S)}. \tag{6.5}$$

However, since the virtual (empty) 2p orbitals of Li have an energy not too far from that of the 2s orbital, it is appropriate to include those orbitals. This means that the σ m.o.s are constructed from the 2s and the $2p_x$ orbitals of both Li atoms, the internuclear axis being chosen as the x axis. Thus, four σ m.o.s are obtained. If there was overlap only between orbitals of the same type, two m.o.s (one bonding and one anti-bonding) would arise from the 2s orbitals, and similarly for the $2p_x$ orbitals. However, there is a net overlap between a 2s orbital of one atom and the $2p_x$ orbital of the other. Thus, all four a.o.s must be considered simultaneously, although in two of them the dominant contribution is 2s whereas in the other two it is 2p, as a consequence of the different energies of 2s and 2p orbitals. The σ m.o. of lowest energy is bonding and has no nodes between the atoms:

$$\sigma_1 = c_1(\varphi_{2sA} + \varphi_{2sB}) + c_2(\varphi_{2xA} + \varphi_{2xB}). \tag{6.6}$$

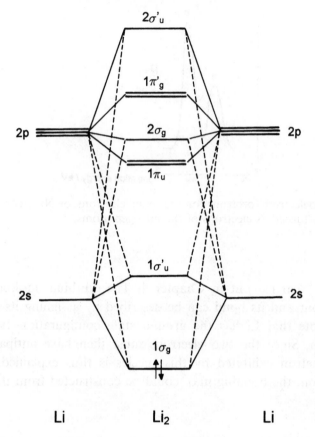

Fig. 6.5 Valence m.o. diagram for Li_2 (full lines and dotted lines refer, respectively, to major and minor contributions of a.o.s to m.o.s).

In this expression, $c_1 > c_2$ in accordance with the dominant contribution of the low-lying 2s atomic orbitals.

There will also be four π m.o.s constructed from the two $2p_y$ and the two $2p_z$ atomic orbitals, as seen on page 80 for excited configurations of H_2^+. They form two degenerate pairs, one bonding and the other anti-bonding. Figure 6.5 shows the qualitative m.o. diagram for Li_2. Reference to the parity of the various m.o.s is included (see page 75); for example, σ_1 is $1\sigma_g$, that is, the function is symmetric with respect to inversion about the centre of the molecule.

The energy difference between the anti-bonding and bonding π m.o.s is smaller than for σ m.o.s. This is partly explained on the basis of the expressions for the energies of π and π' which are similar to those encountered on page 78 for the σ m.o.s of H_2^+: $N(\alpha + \beta)$ and $N'(\alpha - \beta)$, respectively. The absolute value of the resonance integral β is small in view of the small overlap integral for $2p_\pi$ atomic orbitals:

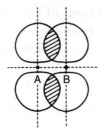

Besides, if those m.o.s were occupied, the increased electron density between the nuclei due to constructive combination of $2p_\pi$ orbitals (bonding m.o.) and the corresponding decreased electron density for the destructive combination (anti-bonding m.o.) would occur mainly for regions on opposite sides of the internuclear axis. Hence, the bonding and anti-bonding effects are smaller than for σ m.o.s.

Another main feature to note in Fig. 6.5 is that, due to a dominant contribution of the low-lying 2s orbitals, the anti-bonding $1\sigma'_u$ lies lower in energy than the bonding m.o.s $1\pi_u$ and $2\sigma_g$ m.o.s.

The ground state configuration is

$$(1s_A)^2, (1s_B)^2, (1\sigma_g)^2$$

and the corresponding isodensity contour plots for a plane containing the bond axis are compared with H_2 in Fig. 6.6.

Due to the appreciable size and diffuse nature of $n=2$ orbitals for atoms of small nuclear charge, the bond is relatively weak: bond energy $105\,\text{kJ}\,\text{mol}^{-1}$ and bond length $267\,\text{pm}$ to be compared with $432\,\text{kJ}\,\text{mol}^{-1}$ and $74\,\text{pm}$ for H_2.

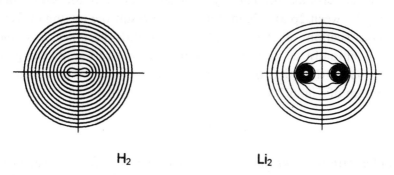

H_2 Li_2

Fig. 6.6 Isodensity contour plots for planes containing the internuclear axes of H_2 and Li_2 (reprinted with permission from ref. 67).

The molecules Be_2, B_2 and C_2

The two additional valence electrons of Be_2 relative to Li_2 (note that Be: $1s^2$, $2s^2$) would lead to a stronger bond should they be assigned to a bonding m.o. However, according to Fig. 6.5 which qualitatively applies to this case, they are anti-bonding:

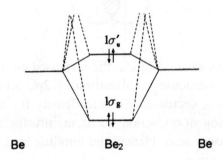

The configuration is

$$(1s_A)^2, (1s_B)^2, (1\sigma_g)^2, (1\sigma_u')^2.$$

The molecule exists at low temperature and with a very small bond energy ($55\,\text{kJ}\,\text{mol}^{-1}$) because the anti-bonding $1\sigma_u'$ is relatively stabilized due to the participation of the $2p_x$ atomic orbitals.

The bond energy increases on going to B_2 ($293\,\text{kJ}\,\text{mol}^{-1}$) which is expected on the basis of two additional bonding electrons. According to the m.o. diagram reproduced in Fig. 6.7, they are assigned to the degenerate $1\pi_u$ bonding m.o.s, one for each m.o. with parallel spins. As a result the molecule is paramagnetic.

As we proceed along the second row in the Periodic Table, the energies of the 2s and 2p orbitals become more negative and, at the same time, the energy gap between 2p and 2s increases. As a result the mixing of 2s and 2p in the construction of the σ molecular orbitals becomes less and less relevant. Eventually, as we shall see below for O_2 and F_2, the energy of $2\sigma_g$ becomes smaller than that of $1\pi_u$ (ref. 68). For C_2, however, essentially the same diagram of Fig. 6.7 applies:

$$(1s_A)^2, (1s_B)^2, (1\sigma_g)^2, (1\sigma_u')^2, (1\pi_u)^4.$$

This configuration shows that there is a net bonding effect equivalent to four electrons. Accordingly, the bond energy is high ($602\,\text{kJ}\,\text{mol}^{-1}$) and the bond is short: 124 pm.

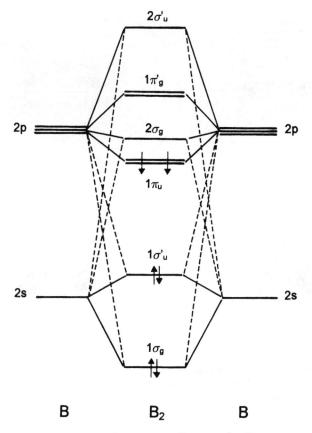

Fig. 6.7 Valence m.o. diagram for B_2.

The molecule N_2

The expected lowest energy configuration for the valence shell is now

$$(1\sigma_g)^2, (1\sigma_u')^2, (1\pi_u)^4, (2\sigma_g)^2 \quad \text{or} \quad (1\sigma_g)^2, (1\sigma_u')^2, (2\sigma_g)^2, (1\pi_u)^4$$

depending on the relative energies of the $2\sigma_g$ and $1\pi_u$ molecular orbitals. In any case, there will be a net bonding effect equivalent to six electrons in agreement with the Lewis structure

$$:N\equiv N: \quad \text{or} \quad |N\equiv N|$$

The *bond order* defined as half the difference between the number of bonding (N_b) and the number of anti-bonding (N_a) electrons

$$\text{Bond order} = (N_b - N_a)/2 \tag{6.7}$$

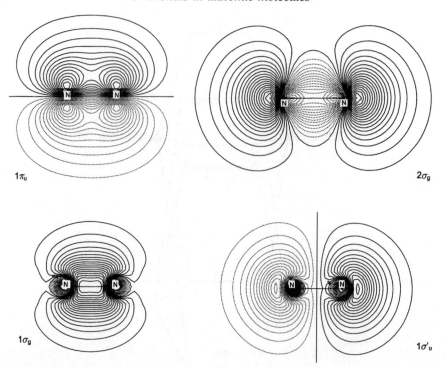

Fig. 6.8 Isocontours for the occupied valence m.o.s of N_2 (only one of the two degenerate π orbitals is shown). Full and dotted lines correspond to opposite signs of the m.o. functions.

is three. The bond energy is high ($942\,kJ\,mol^{-1}$) and the bond is short: $110\,pm$.

Figure 6.8 shows successive contours corresponding to equal values for the occupied m.o.s of N_2 (as calculated using the SCF method mentioned on page 163 and which can be compared with an early publication by Jorgensen and Salem, ref. 69).

It is found that the energies of $2\sigma_g$ and $1\pi_u$ are very close, the $2\sigma_g$ m.o. lying lower in energy according to accurate calculations. However, for N_2^+ this order is reversed, due to orbital relaxation upon ionization of N_2.

Problem 6.2 Give the electronic configuration of lowest energy for N_2^+ and explain why the bond is weaker than in N_2.

Such an orbital reorganization precludes a direct connection being established between the electronic configuration of lowest energy for N_2 and the photoelectron spectrum which is shown in Fig. 6.9; that is, there is breakdown in Koopmans' theorem. For example, the first band (ionization energy around $15\,eV$) is attributed to the removal of one electron from $2\sigma_g$,

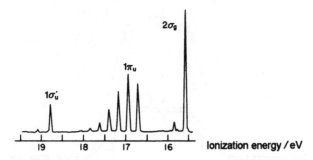

Fig. 6.9 UV-photoelectron spectrum of N_2 (using incident photons of 21.2 eV).

whereas the central band of Fig. 6.9 is assigned to the removal of one $1\pi_u$ electron. This is contrary to expectation from the point of view of the N_2 configuration, but is in agreement with the N_2^+ lowest energy configuration which is $(1s_A)^2$, $(1s_B)^2$, $(1\sigma_g)^2$, $(1\sigma_u')^2$, $(1\pi_u)^4$, $(2\sigma_g)^1$ and not $(1s_A)^2$, $(1s_B)^2$, $(1\sigma_g)^2$, $(1\sigma_u')^2$, $(2\sigma_g)^2$, $(1\pi_u)^3$. A detailed analysis of Koopmans' theorem in the case of molecules is given in ref. 70.

The central band in Fig. 6.9 is split, as a result of part of the energy of the incident photon being used in various vibrational excitations of the produced ion, N_2^+. This type of splitting is characteristic of removal of one electron which significantly affects the forces acting on the nuclei.

The molecules O_2 *and* F_2

In the photoelectron spectrum of O_2, the bands corresponding to $1\pi_u$ and $2\sigma_g$ change positions relative to each other (Fig. 6.10). This agrees with $2\sigma_g$ lying substantially below $1\pi_u$, as found in quantum-mechanical calculations.

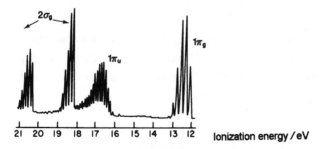

Fig. 6.10 Photoelectron spectrum of O_2 (using incident photons of 21.2 eV) (adapted from ref. 66).

The m.o. diagram for O_2 is shown in Fig. 6.11. There are some other differences when comparing with N_2. First, the two additional electrons

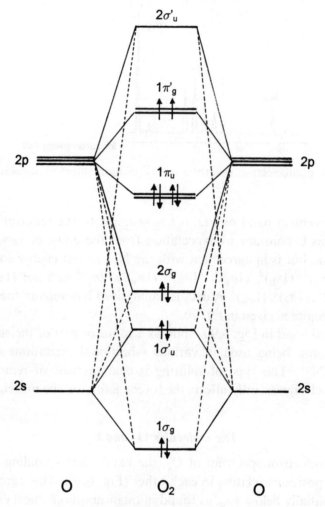

Fig. 6.11 Valence m.o. diagram for O_2.

are attributed to anti-bonding m.o.s $(1\pi'_g)$ thus weakening the bond. This implies a bond order of 2 and a smaller bond energy, in agreement with observation (494 kJ mol^{-1}). We now have a double bond

$$\overline{O} = \overline{O}$$

On the other hand, two electrons assigned to two degenerate m.o.s lead to a ground state having one electron in each m.o. with parallel spins. As a result, the molecule is paramagnetic.

This magnetic feature of O_2 is the reason for the occurrence of two split bands corresponding to $2\sigma_g$; an ionization from this orbital leaves one electron whose spin can be either parallel or antiparallel with respect to the electron spins of $1\pi_u$ leading to two electronic states of O_2^+.

Problem 6.3 Write the electronic configurations of lowest energy for O_2^- and O_2^{2-}.

For the molecule F_2, the electronic configuration is

$$(1s_A)^2, (1s_B)^2, (1\sigma_g)^2, (1\sigma_u')^2, (2\sigma_g)^2, (1\pi_u)^4, (1\pi_g')^4$$

the bond order is 1 and the bond is weak (bond energy of $155\,kJ\,mol^{-1}$) partly as a result of the π anti-bonding m.o.s being more anti-bonding than the π bonding m.o.s are bonding. It is this same argument that explains the non-existence of diatomic molecules of neon, whose configuration would be

$$(1s_A)^2, (1s_B)^2, (1\sigma_g)^2, (1\sigma_u')^2, (2\sigma_g)^2, (1\pi_u)^4, (1\pi_g')^4, (2\sigma_u')^2.$$

6.4 Heterodiatomic molecules

For a heteronuclear diatomic molecule, in general the electrons are not shared equally by the two atoms. First, the number of electrons contributed by each atom is not the same. Secondly, the electrons effectively involved in bonding do not usually have a symmetric distribution about the two atoms. As a result, heteronuclear diatomic molecules are usually polar. In the simple case of H–X bonds, the polarity of the HX molecule is directly related to the *electronegativity* difference of the two atoms. For example, HF is strongly polar in the sense

$$\delta+ \quad H-F \quad \delta-$$

whereas LiH is polar in the sense

$$\delta+ \quad Li-H \quad \delta-$$

in accordance with the following sequence of electronegativities

$$\chi_F > \chi_H > \chi_{Li}.$$

The polarity of these bonds has a counterpart in the form of the bonding molecular orbitals. Let us assume that, in a molecule H_2, atom B would become more electronegative

$$H-H$$
$$\delta+ \quad A \quad B \quad \delta-$$

(eventually by acquiring an additional proton, thus becoming a He nucleus). The result would be a higher electron density in B than in A. In m.o. terms, the 1s orbital of B would lie lower in energy

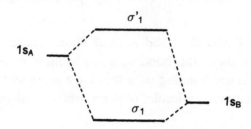

and, thus, its contribution to the bonding m.o. would be increased:

$$\sigma_1 = c_A\varphi_{1sA} + c_B\varphi_{1sB} \tag{6.8}$$

with

$$c_B > c_A.$$

The probability density function would be

$$\sigma_1^2 = c_A^2\varphi_{1sA}^2 + c_B^2\varphi_{1sB}^2 + 2c_ac_B\varphi_{1sA}\varphi_{1sB} \tag{6.9}$$

leading to higher values in the vicinity of B relative to those near A.

Meanwhile, the corresponding anti-bonding m.o. would be

$$\sigma_1' = c_A'\varphi_{1sA} - c_B'\varphi_{1sB} \tag{6.10}$$

with

$$c_A' > c_B'$$

to warrant orthogonality.

For the hypothetical example above, if the difference of electronegativities becomes very large, there will be a greater difference between the energies of $1s_A$ and $1s_B$, and

$$c_B \gg c_A$$

the result being

$$\sigma_1 \cong \varphi_{1sB}.$$

With the two electrons described only by φ_{1sB}, an ionic bond would arise:

$$A^+ \quad B^-$$

Let us consider now a few realistic examples, beginning with HF which has a dipole moment of 1.91 D. In a first approximation, we neglect the contribution of the 2s orbital of F to the bonding, because of its very low energy compared with both the 2p orbitals of F and the 1s orbital of H. We then have:

$$\sigma_1 = c_H \varphi_{1sH} + c_F \varphi_{2xF} \tag{6.11}$$

$$\sigma_1' = c_H' \varphi_{1sH} - c_F' \varphi_{2xF} \tag{6.12}$$

if the internuclear axis is named the x axis. The coefficients can be determined by an application of the variation principle (see Chapter 7):

$$\sigma_1 = 0.33\, \varphi_{1sH} + 0.94\, \varphi_{2xF} \tag{6.13}$$

$$\sigma_1' = 0.94\, \varphi_{1sH} - 0.33\, \varphi_{2xF}. \tag{6.14}$$

Problem 6.4   Confirm that the m.o.s above are orthogonal.

Since the $2p_y$ and $2p_z$ orbitals of F are orthogonal with the 1s orbital of H, they are not altered and become filled non-bonding orbitals in HF, in addition to the filled non-bonding 2s orbital of F (Fig. 6.12(a)). Hence, the formula

$$H-\overline{\underline{F}}|$$

In a more rigorous treatment, where the 2s orbital of F contributes slightly to bonding, the m.o. diagram is that shown in Fig. 6.12(b).

The relative contributions of the 2s and 2p orbitals of F in the various m.o.s can be recognized in the graphical representations of Fig. 6.13.

For LiH, with a dipole moment of 5.83 D, the proximity of the 2s and 2p orbital energies of Li requires that the 2s orbital is involved in the construction of the σ molecular orbitals. A variation calculation leads to

$$\sigma = 0.685\, \varphi_{1sH} + 0.323\, \varphi_{2sLi} + 0.231\, \varphi_{2xLi}. \tag{6.15}$$

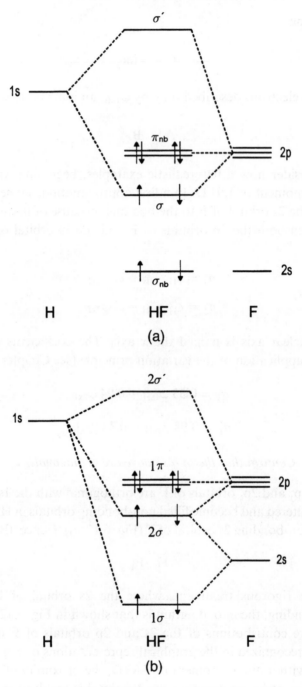

Fig. 6.12 Valence m.o. diagram for HF: (a) considering 2s of F as a core orbital; (b) considering the small participation of 2s of F in bonding.

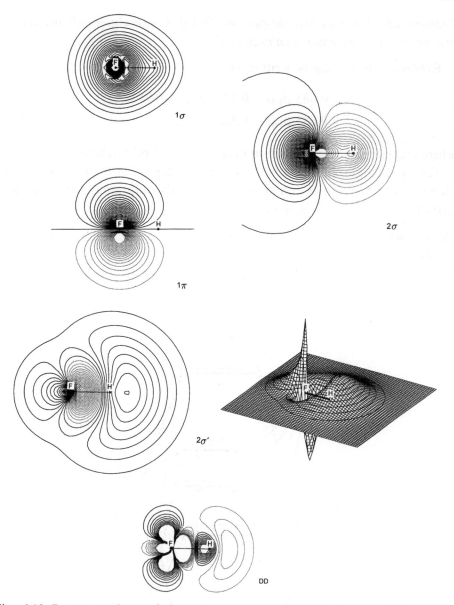

Fig. 6.13 Representations of the valence m.o.s of HF (only one of the two π orbitals is shown) by contour plots and a three-dimensional grid in the case of the vacant (anti-bonding) m.o. Plot of DD (density difference) illustrates the detailed structure of the difference map between the total electron density and the atomic contributions if no bond was formed (full lines for increase of electron density and dotted lines for decrease).

Problem 6.5   Confirm that the function (6.15) is normalized only because the overlap integrals have not been neglected.

Expression (6.15) can be written as

$$\sigma = 0.685\,\varphi_{1sH} + 0.323\,(\varphi_{2sLi} + 0.715\varphi_{2xLi})$$
$$= 0.685\,\varphi_{1sH} + 0.323\,\varphi_{(sp)Li} \tag{6.16}$$

where $\varphi_{(sp)Li}$, replacing the sum in brackets, is a hybrid orbital of Li.

The m.o. diagram for CO is more complex than that for HF. Now, both atoms contribute with 2s and 2p orbitals, and the diagram is more similar to that of N_2 of which CO is, in fact, isoelectronic (Fig. 6.14).

Problem 6.6   By analogy with Fig. 6.9, predict the UV-photoelectron spectrum of CO.

Problem 6.7   Construct a m.o. diagram for CN⁻ *and write its electronic configuration.*

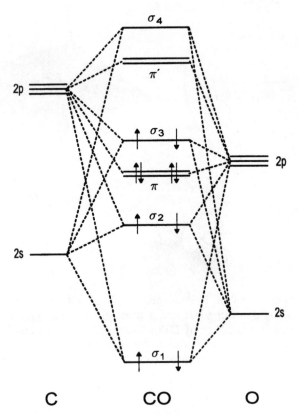

Fig. 6.14 Valence m.o. diagram for CO. The designation g and u is dropped because of the lack of a centre of symmetry.

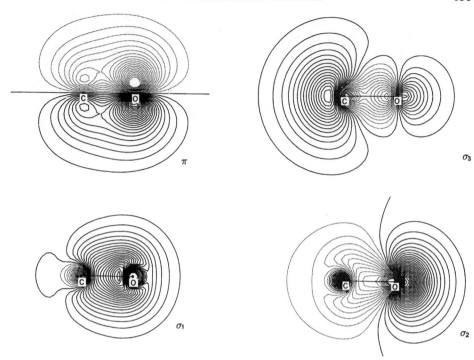

Fig. 6.15 Isocontours for the occupied valence m.o.s of CO (only one of the two degenerate π orbitals is shown). Full and dotted lines correspond to opposite signs of the m.o. functions.

Figure 6.15 shows the isocontours for the occupied m.o.s of CO which can be compared with Fig. 6.8 (see also ref. 69).

The main contribution to σ_2 comes from the 2s and $2p_x$ orbitals of C; it is almost a non-bonding orbital localized in the carbon atom. This orbital is involved in bonding interactions with various transition metal orbitals in coordination chemistry. The charge distribution in σ_2 partially compensates the polarization associated with the other occupied m.o.s in the sense predicted by the electronegativity difference:

$$\delta+ \quad C \quad O \quad \delta-$$

This factor, together with a significant polarization of the atomic densities, particularly that of the carbon atom, opposing the pronounced charge transfer related to the different electronegativities (ref. 71), explains the close-to-zero dipole moment of CO (0.10 D). In Fig. 6.16 a comparison is drawn between the isodensity contours of CO and N_2 for a plane containing the internuclear line.

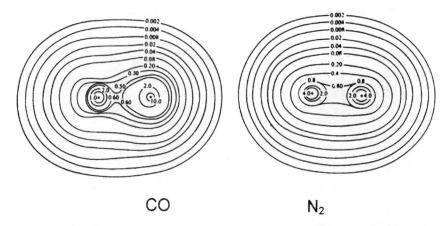

CO N_2

Fig. 6.16 Isodensity contours of CO and N_2 for a plane containing the inter-nuclear line (ref. 67).

For recent reviews on Hartree–Fock and density-functional calculations in diatomic molecules, see refs. 61 and 72.

6.5 Electronegativity

The concept of electronegativity was originally introduced in 1932 by the American chemist Linus C. Pauling (1901–1994, 1954 Nobel laureate in Chemistry) as the power of an atom in a molecule to attract electrons to itself (ref. 73). The Pauling scale of electronegativities, χ_P, is based on the general observation that the polarity of a diatomic molecule AB is accompanied by an additional stabilization. In particular, the bond dissocia-tion energy of AB (D_{A-B}) is higher than the mean of the bond dissociation energies of A_2 (D_{A-A}) and B_2 (D_{B-B}). Thus, through yet another combina-tion of quantum-mechanical and thermodynamical arguments, a scale is constructed in terms of the expression:

$$96.5(\chi_{P,A} - \chi_{P,B})^2 = D_{A-B} - (D_{A-A}D_{B-B})^{1/2}. \qquad (6.17)$$

According to this definition (see Chapter 10 for bond energies), the Pauling electronegativities range from 0.79 for Cs to 3.98 for F, being 2.20 for H, 2.55 for C, 3.04 for N and 3.44 for O.

An important step towards an evaluation of the quantity from first principles was taken two years later by another American chemist, Robert S. Mulliken (1896–1986, 1966 Nobel laureate in Chemistry), yielding a pro-portionality between electronegativity, χ_M, and the arithmetic mean of the ionization energy, I, and the electron affinity, E_{ea}, of the atom (ref. 74):

$$\chi_M \propto (I + E_{ea})/2. \tag{6.18}$$

To a first approximation, I is the symmetrical of the energy of the highest occupied orbital whereas E_{ea} is the symmetrical of the energy of the lowest unoccupied orbital. In this scale, some electronegativities to be compared with the Pauling values are: 3.06 for H, 2.67 for C, 3.08 for N, 3.21 for O and 4.42 for F. For Ne, a value $\chi_M = 4.60$ is obtained, there being no counterpart in the Pauling scale due to the chemical inertness of this noble gas.

Other scales have been proposed, namely by Allred and Rochow, in 1958 (ref. 75), and by Allen, in 1989 (ref. 76). In the first case, the electronegativity, χ_{AR}, is an atomic property taken proportional to the force exerted by the effective nuclear charge at the periphery of the atom:

$$\chi_{AR} \propto [(3590\, Z_{eff}/r^2) + 0.744] \tag{6.19}$$

where r is the covalent radius of the atom expressed in pm. Elements with the highest electronegativity on the Allred–Rochow scale are those, like fluorine, with the highest effective nuclear charges and the smallest radii. They are also the elements of higher ionization energies and electron affinities. Some values to be compared with the Pauling electronegativities are: 2.20 for H, 2.50 for C, 3.07 for N, 3.50 for O and 4.10 for F.

The electronegativity scale proposed by Allen is defined in terms of the average energy of the electrons in the valence s and p orbitals of an atom:

$$\chi_A \propto (n_s \varepsilon_s + n_p \varepsilon_p)/(n_s + n_p) \tag{6.20}$$

where n_s and n_p are the number of electrons in the s and p orbitals and the orbital energies ε are obtained from averaged spectroscopic data. In general, atoms having valence orbitals of lower energy also have higher I and E_{ea} values. Some appropriate values to be compared with the Pauling electronegativities are: 2.30 for H, 2.54 for C, 3.07 for N, 3.61 for O and 4.19 for F. For Ne, a value of 4.79 is obtained.

It has also been shown that the mean of the ionization energy and the electron affinity, as the basis of the Mulliken scale, is, to a good approximation, equal to the slope of the electronic energy E with respect to the addition of electrons. Thus, yet another plausible definition of electronegativity is to identity the electronegativity of an atom with the negative of the so-called *electronic chemical potential*:

$$\mu = -(\partial E/\partial N) \tag{6.21}$$

where N is the number of electrons. In a molecule, electrons will tend to flow towards regions of higher electronegativity which are regions of lower electronic chemical potential. In this way, the electronegativities of the atoms forming a molecule will tend to equalize and acquire the same uniform value. Expression (6.21) appears in a natural way within the framework of density-functional theory along the lines defined by Parr and Yang (ref. 77). For a brief account of the different definitions of electronegativity, see ref. 78.

7

Orbitals in polyatomic molecules

7.1 New features relative to diatomic molecules

The existence of more than two nuclei in polyatomic molecules means that
molecular orbitals are polycentric, extending, in principle, over the whole
molecule. It is in this sense that molecular orbitals in polyatomic systems are
said to be *delocalized*. Within the linear combination of atomic orbitals
(l.c.a.o.) approach, valence m.o.s are approximated by linear combinations
of the valence a.o.s of all the intervening atoms. Just as for diatomic molecules, this is an approximation because only a finite set of basis functions
are being used. The eventual inclusion of virtual a.o.s of energy close to the
valence orbital energies can help to improve the results.

From a certain number, M, of a.o.s φ_n an equal number of m.o.s ψ_k are
constructed (conservation of orbital number):

$$\psi_k = \sum_1^M c_n \varphi_n \quad (k = 1, 2, \ldots, M) \ (n = 1, 2, \ldots, M). \qquad (7.1)$$

The electronic wavefunction Ψ is a Slater determinant (or a combination of
Slater determinants in the case of open-shell systems) based on the occupied
m.o.s, each one taken twice (if doubly occupied) and associated either with
α or β spin functions (molecular spin-orbitals)

$$\Psi = N \begin{vmatrix} \psi_1(1) & \psi_1(2) & \psi_1(3) & \cdots \\ \bar{\psi}_1(1) & \bar{\psi}_1(2) & \bar{\psi}_1(3) & \cdots \\ \psi_2(1) & \psi_2(2) & & \cdots \\ \bar{\psi}_2(1) & \bar{\psi}_2(2) & & \cdots \\ \cdots \end{vmatrix} . \qquad (7.2)$$

This can be simply represented by the product of the diagonal functions
between short vertical bars: $|\psi_1(1) \, \bar{\psi}_1(2) \, \psi_2(3) \ldots|$.

The total electron density function (for real functions) is given by

$$\Psi^2 = 2 \sum_{occ.} \psi_k^2 \tag{7.3}$$

for doubly occupied orthogonal m.o.s.

The delocalized picture of m.o.s for polyatomic molecules requires some adaptation of the distinction between σ and π m.o.s, particularly for non-linear molecules. If π m.o.s continue to be defined as linear combinations of p orbitals having parallel axes, σ m.o.s are to be regarded as the remaining m.o.s, even though, in general, they no longer have internuclear lines as axes of cylindrical symmetry. In Chapter 8, we will see how to reconcile delo-calized m.o.s with the classical structural formulae involving bonds between adjacent atoms.

As in the previous chapter for diatomics, we shall now develop an essentially qualitative m.o. description of the structure of small, highly symmetric polyatomic molecules. This will be followed by a brief presenta-tion of the quantitative rationale for any molecule. Besides relying on the symmetry of the systems, we will often make use of the rule that the smaller the number of nodes (nodal surfaces) between adjacent atoms for each m.o., the greater the bonding and, hence, the lower its energy. A node (of radial origin, and not of angular origin as is the case for p orbitals) coinciding with the centre of an atom occurs when there is no contribution from that atom to the m.o. under consideration.

Another novel situation for polyatomic molecules relative to diatomics is the existence of bond angles. We shall see that qualitative m.o. theory can also be of use in the understanding of some molecular geometries.

7.2 Molecular orbitals in AH_n molecules

In the same way that the study of H_2^+ and H_2 served as a basis for the construction of m.o.s in other diatomic molecules, the discussion of the simplest triatomic species H_3^+ will greatly help the study of more complex molecules of the general formula AH_n.

The H_3^+ ion

This unstable species was first detected in the gaseous phase by mass spectrometry when H_2 was subjected to an electric discharge. It is known from spectroscopic studies that it has an equilateral geometry with bond lengths around 100 pm.

Let us begin, however, by considering a linear arrangement of three H atoms, each contributing a 1s orbital φ_{1s}

There will be three σ m.o.s. That of lowest energy is equally bonding between A and B and between B and C and is given by:

$$\sigma_1 = (\varphi_{1sA} + \mu\varphi_{1sB} + \varphi_{1sC})N_1. \tag{7.4}$$

The normalization factor N_1 is included and μ is a numerical coefficient which accounts for the fact that B is in a different situation when compared to A and C. The shapes and signs of the contributing atomic orbitals are shown below Eq. (7.4).

The next in energy, σ_2, must have one node (one nodal surface). The symmetry of the system requires that such a node is located in the central atom B. Thus, φ_{1sB} will not contribute to σ_2:

$$\sigma_2 = (\varphi_{1sA} - \varphi_{1sC})N_2. \tag{7.5}$$

Since A and C are not directly connected to each other, the interaction between the respective 1s orbitals is small, and σ_2 is essentially a non-bonding m.o. (slightly anti-bonding).

Problem 7.1 Using the expression $S = e^{-r}(1 + r + r^2/3)$ for the overlap integral between two hydrogen 1s orbitals (ref. 79) (with r in atomic units, 1 atomic unit or Bohr radius being 52.9 pm), compare the overlap integrals for $r = 2$ and $r = 3$.

The third m.o. will be anti-bonding both between A and B and between B and C; i.e. with nodes between A and B and between B and C:

$$\sigma_3 = (\varphi_{1sA} - \lambda\varphi_{1sB} + \varphi_{1sC})N_3. \tag{7.6}$$

Problem 7.2   Confirm that σ_2 and σ_3 are orthogonal by calculating the integral $S_{2,3} = \int \sigma_2 \sigma_3 \, dv$.

As was performed for H_2^+, we can obtain expressions for the energies of the three m.o.s as a function of the following integrals, the coulomb integral α and the resonance integral β,

$$\alpha = \langle s_A|\hat{H}|s_A\rangle = \langle s_B|\hat{H}|s_B\rangle = \langle s_C|\hat{H}|s_C\rangle$$
$$\beta = \langle s_A|\hat{H}|s_B\rangle = \langle s_B|\hat{H}|s_C\rangle \tag{7.7}$$

and considering

$$\langle s_A|\hat{H}|s_C\rangle = 0. \tag{7.8}$$

In the integrals above, s_X represents φ_{1sX}. For E_1 we get

$$\begin{aligned} E_1 &= \langle \sigma_1|\hat{H}|\sigma_1\rangle = N_1^2[\langle s_A|\hat{H}|s_A\rangle + \mu^2\langle s_B|\hat{H}|s_B\rangle + \langle s_C|\hat{H}|s_C\rangle \\ &\quad + 2\mu\langle s_A|\hat{H}|s_B\rangle + 2\mu\langle s_B|\hat{H}|s_C\rangle] \\ &= N_1^2[(2+\mu^2)\alpha + 4\mu\beta] \end{aligned} \tag{7.9}$$

and, similarly,

$$E_2 = N_2^2(2\alpha)$$
$$E_3 = N_3^2[(2+\lambda^2)\alpha - 4\lambda\beta]. \tag{7.10}$$

When an atom C is placed at equal distances from A and B, in an equilateral geometry,

all atomic orbitals are to be considered equally: $\mu=1$ and, to warrant orthogonality, $\lambda=-2$.

Problem 7.3   Prove that, for the equilateral arrangement, $\sigma_1 = (\varphi_{1sA} + \varphi_{1sB} + \varphi_{1sC})N_1$ and $\sigma_3 = (\varphi_{1sA} - 2\varphi_{1sB} + \varphi_{1sC})N_3$ are orthogonal.

Since, in addition,

$$\langle s_A|\hat{H}|s_C\rangle = \beta \tag{7.11}$$

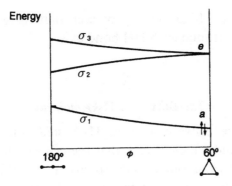

Fig. 7.1 Energy of the m.o.s for three symmetrically arranged H atoms as a function of the HHH angle.

a calculation similar to Eq. (7.10) yields

$$E_1 = N_1^2(3\alpha + 6\beta)$$
$$E_2 = N_2^2(2\alpha - 2\beta)$$
$$E_3 = N_3^2(6\alpha - 6\beta). \tag{7.12}$$

Problem 7.4 *By using the appropriate values for* N_2 *and* N_3, *prove that* $E_2 = E_3$.

Molecular orbitals σ_2 and σ_3 become degenerate. According to the nomenclature of group theory, they are named 'e' functions (doubly degenerate) whereas the non-degenerate σ_1 m.o. is classified as 'a'.

Figure 7.1 shows the energy change for each m.o. as the ABC angle decreases from 180° to 60°. The stabilization of σ_1 and σ_3 results from increasing bonding involving A and C, whereas the increase of σ_2 energy is a result of the increased anti-bonding interaction of A and C.

The two electrons of H_3^+ are assigned to the orbital σ_1, with the triangular geometry corresponding to a lower energy for this system than for a linear arrangement. Figure 7.2 shows the isodensity contours in the

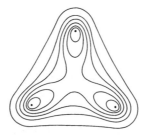

Fig. 7.2 Isodensity contours in the nuclear plane of H_3^+.

nuclear plane of H_3^+. It can clearly be seen that there is no symmetry of the electron density around each HH bond axis.

The BeH₂ and H₂O molecules

The main difference between a linear H_3 system and beryllium hydride, BeH_2, which is a linear molecule, is that the central atom now contributes two valence electrons and four atomic orbitals: the 2s orbital, which is doubly occupied in an isolated ground state Be atom (configuration: $1s^2, 2s^2$), and the three 2p orbitals which are empty in Be but lie not too far above 2s in energy. Graphically, we have:

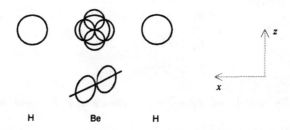

The $2p_y$ and $2p_z$ orbitals of Be (the x axis is chosen as the internuclear line) are both orthogonal to the 1s orbitals of the H atoms; hence, they are non-bonding orbitals in BeH_2. The remaining four a.o.s give rise to four σ m.o.s, the forms of which can be established as follows.

The m.o. of lowest energy will be bonding both in section H(1)Be and in BeH(2) and will have no nodes (except, perhaps, for a nodal surface associated with the 2s orbital of Be). Thus, the $2p_x$ orbital of Be cannot participate in this m.o. and the only possible combination is:

$$\sigma_1 = s_1 + as + s_2. \tag{7.13}$$

In this expression, for simplicity, s_i represents $\varphi_{1sH(i)}$ whereas s represents φ_{2sBe}. The contribution of the latter is a and, for symmetry reasons, the contribution of the H atoms are the same. The normalization factor was ignored.

The next m.o. will have a node on the central atom. This means that the 2s orbital of Be is not involved, but the $2p_x$ orbital (represented by x) participates:

$$\sigma_2 = s_1 + bx - s_2. \qquad (7.14)$$

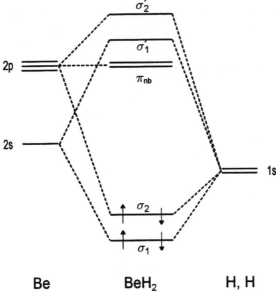

With two nodes, we have the following m.o. which is anti-bonding in each HBe fragment:

$$\sigma_1' = s_1 - a's + s_2. \qquad (7.15)$$

The last m.o. has three nodes, two on the HBe fragments and one on the central atom:

$$\sigma_2' = -s_1 + b'x + s_2. \qquad (7.16)$$

Figure 7.3 shows the m.o. diagram including the π non-bonding orbitals: $2p_y$ and $2p_z$ of Be. The four electrons occupy the bonding m.o.s σ_1 and σ_2:

$$(\sigma_1)^2, (\sigma_2)^2.$$

Fig. 7.3 Valence m.o. diagram for BeH$_2$.

The contours (in a plane containing the nuclear axis) corresponding to equal values of each of the delocalized functions σ_1 and σ_2 are plotted in Fig. 7.4. It can be noted that the two formal bonds HBe are described in the same manner. This means that the four bonding electrons are equivalent to an average of two electrons per bond (single bonds):

$$H-Be-H.$$

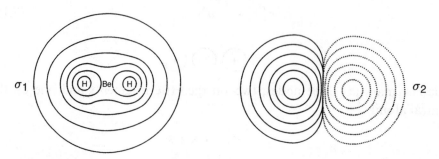

Fig. 7.4 Contours for equal values of each of the bonding m.o.s of BeH$_2$.

Problem 7.5 Confirm that the total probability density $\sigma^2 = \sigma_1^2 + \sigma_2^2$ is symmetric with respect to the Be nucleus.

The m.o. diagram of Fig. 7.3 would qualitatively apply to H$_2$O if this molecule was linear, with eight valence electrons being considered instead of four. The discussion that follows is centred on the changes of the various m.o.s when the HOH angle decreases from a linear arrangement to the actual geometry of the water molecule.

For the following choice of axes

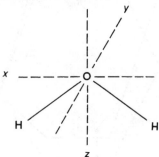

the 2p$_y$ orbital of O remains non-bonding: π_y. The molecular orbital σ_2 has the same expression as for BeH$_2$ (or a linear H$_2$O), although it will become less bonding because of the smaller overlap between the hydrogen 1s orbitals and the oxygen 2p$_x$ orbital as the angle decreases:

Figure 7.5 shows the way that the energy of σ_2 changes with the bond angle.

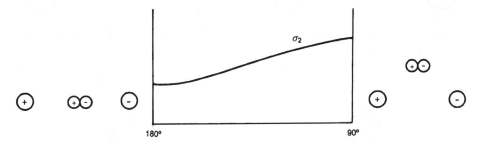

Fig. 7.5 Energy of m.o. σ_2 for AH_2 as a function of the HAH angle.

Similarly, the corresponding anti-bonding m.o. σ_2' has essentially the same form:

The remaining three m.o.s require a little more attention. The reason is that $2p_z$ is no longer a non-bonding orbital; it has a non-zero overlap with the hydrogen s orbitals. Thus, the m.o. of lowest energy is altered by constructive interference from $2p_z$ (represented by z) and becomes

$$\sigma_1 = s_1 + as + cz + s_2 \tag{7.17}$$

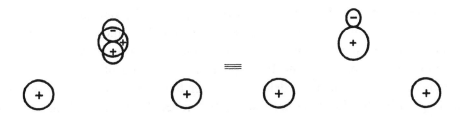

whereas the corresponding anti-bonding m.o. is

$$\sigma_1' = s_1 - a's - c'z + s_2. \qquad (7.18)$$

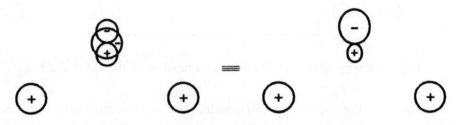

Finally, instead of $\pi_z = 2p_z$, a weakly bonding m.o. (or, effectively, almost non-bonding) arises:

$$\psi_z = s_1 - ds + ez + s_2. \qquad (7.19)$$

In the graphical representations above, the joint contribution of the oxygen 2s and $2p_z$ orbitals is presented as a hybrid orbital.

The energies of σ_1 and ψ_z change with the angle as plotted in Fig. 7.6. It is found that it is mainly the decrease in the energy of ψ_z (a 2p non-bonding orbital for the linear geometry) when the HOH angle decreases that is responsible for the bent geometry for H_2O. The plots in Figs. 7.5 and 7.6 are known as *Walsh diagrams*.

The m.o. diagram for H_2O is presented in Fig. 7.7 and the electronic configuration is

$$(\sigma_1)^2, (\sigma_2)^2, (\psi_z)^2, (\pi_y)^2$$

with four bonding electrons, two non-bonding electrons and two practically non-bonding electrons.

These results can be compared with data from the photoelectron spectrum of H_2O shown in Fig. 7.8. It is noted that the removal of a bonding electron affects the vibrational states of the molecule, as expected.

With the exception of the two π_y electrons, all the other electrons are assigned to m.o.s which are delocalized over the three atoms. However, the description of the two OH sections of the molecule is identical, as seen in

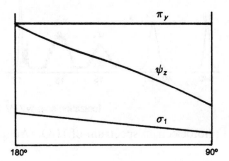

Fig. 7.6 Energies of m.o. σ_1 and ψ_z for AH$_2$ as a function of the HAH angle.

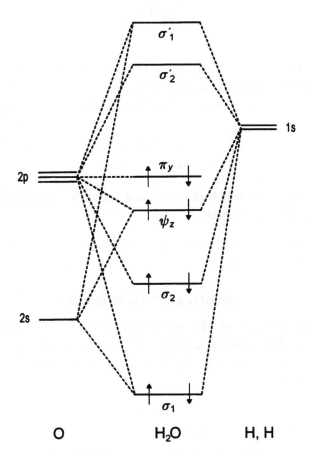

Fig. 7.7 Valence m.o. diagram for H$_2$O.

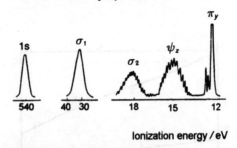

Fig. 7.8 Complete photoelectron spectrum of H_2O. (Adapted from ref. 66.)

Fig. 7.9, in agreement with the classical formula of two equivalent bonds:

An alternative way of discussing the general form of the molecular orbitals in H_2O and BeH_2 (and, similarly, in H_3^+) consists of first associating the hydrogen 1s orbitals in two intermediate linear combinations of different symmetry:

$$\phi_s = s_1 + s_2 \tag{7.20}$$
$$\phi_a = s_1 - s_2 \tag{7.21}$$

and, then, taking these functions as a basis for combination with the 2s and 2p orbitals of the central atom.

Problem 7.6   Show that ϕ_s, but not ϕ_a, is orthogonal to $2p_x$ for a linear AH_2, hence only ϕ_a interacts with $2p_x$.

The BH_3 and NH_3 molecules

The approach outlined in the last paragraphs can be extended to the trigonal planar BH_3 and the pyramidal NH_3 molecules. The intermediate linear combinations of the three hydrogen 1s orbitals have the same form as the m.o.s of H_3^+:

$$\phi_1 = s_1 + s_2 + s_3$$
$$\phi_2 = s_1 - s_2$$
$$\phi_3 = s_1 - 2s_2 + s_3. \tag{7.22}$$

They must be combined with 2s, $2p_x$ and $2p_y$ of the boron atom of BH_3 (the x and y axes chosen to lie in the nuclear plane), leaving the $2p_z$ atomic orbital, orthogonal to all the others, as a non-bonding m.o.

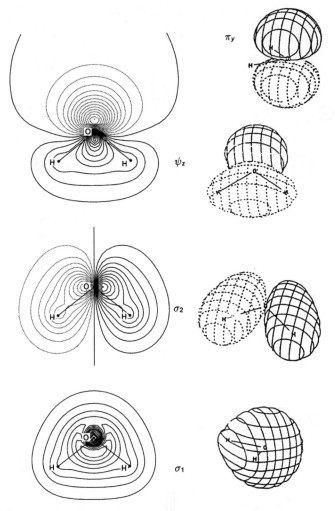

Fig. 7.9 Isocontour plots in the plane of the nuclei for the occupied valence m.o.s of H_2O (that plane is a nodal plane for the π_z orbital) and comparison with the three-dimensional representations of ref. 69.

It can easily be seen that, for example, ϕ_1 is orthogonal to $2p_x$ and $2p_y$

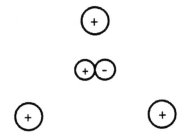

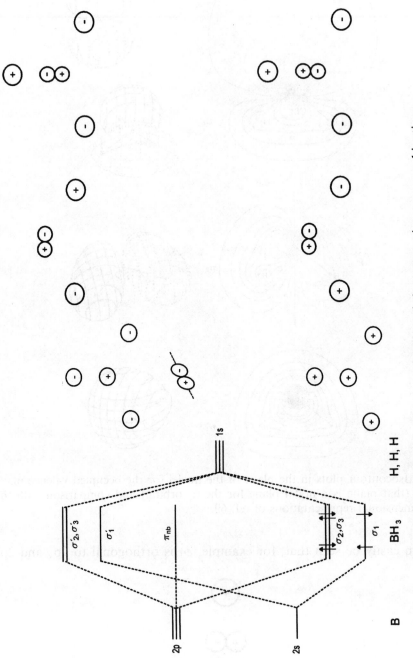

Fig. 7.10 Valence m.o. diagram of BH_3 and the appropriate a.o. combinations.

and that ϕ_2 is orthogonal to 2s and $2p_y$ but not to $2p_x$:

The m.o. diagram is shown in Fig. 7.10, which also depicts the various combinations of a.o.s in the formation of m.o.s. The non-bonding π_z orbital is empty, which explains the Lewis acid character of BH_3. There are six bonding electrons shared by all the atoms, giving an average of two electrons per formal BH bond:

$$\underset{\underset{\displaystyle H}{\overset{\displaystyle |}{\overset{\displaystyle B}{\diagup\diagdown}}}}{H\qquad\qquad H}$$

Molecular orbitals σ_2 and σ_3 are degenerate. This is a consequence of ϕ_2 and ϕ_3 being degenerate (see discussion of H_3^+) in association with the equality of the overlap integrals $\langle\phi_2|2p_x\rangle$ and $\langle\phi_3|2p_y\rangle$.

To go from the m.o.s of BH_3 to the m.o.s of NH_3 we will follow a procedure similar to that we used on going from BeH_2 to H_2O. The result is shown in Fig. 7.11.

Problem 7.7   Compare the form of σ_2 in NH_3 and BH_3 with σ_2 in BeH_2 and H_2O.

Since the σ_{nb} orbital is almost non-bonding, there are six bonding electrons in the molecule in agreement with the Lewis formula:

$$\underset{\underset{\displaystyle H}{\overset{\displaystyle |}{}}}{\overset{\displaystyle \overline{N}}{H\diagup\ \ \diagdown H}}$$

The two electrons in the non-bonding orbital make NH_3 a Lewis base.

The CH_4 molecule

We can arrive at the form of the eight valence m.o.s of CH_4 in a similar fashion to that followed previously. The intermediate combinations of the hydrogen 1s orbitals are obtained by considering the various pairing schemes:

$$\phi_1 = (s_1 + s_2) + (s_3 + s_4)$$

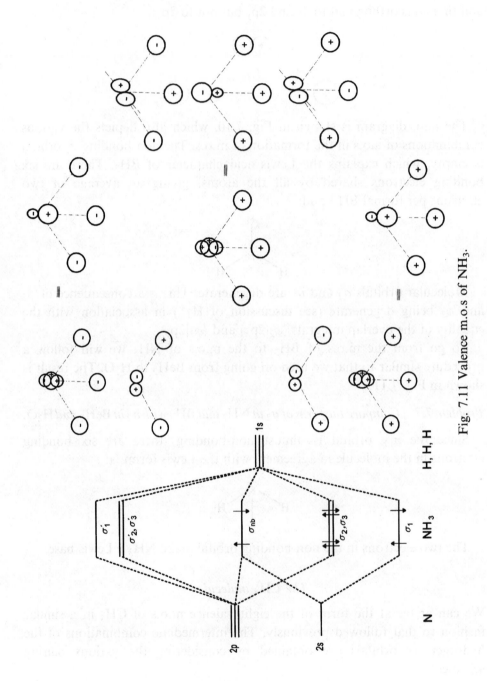

Fig. 7.11 Valence m.o.s of NH₃.

$$\phi_2 = (s_1 + s_2) - (s_3 + s_4)$$

$$\phi_3 = (s_1 - s_2) + (s_3 - s_4)$$

$$\phi_4 = (s_1 - s_2) - (s_3 - s_4). \qquad (7.23)$$

The lowest energy m.o. is bonding everywhere (no nodes):

$$\sigma_1 = \phi_1 + as$$

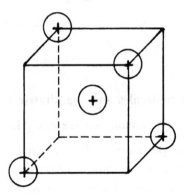

where s represents the 2s carbon atomic orbital. The next three m.o.s are degenerate (compare with σ_2 and σ_3 of BH$_3$ and NH$_3$). The m.o. diagram is given in Fig. 7.12. All the eight electrons are bonding.

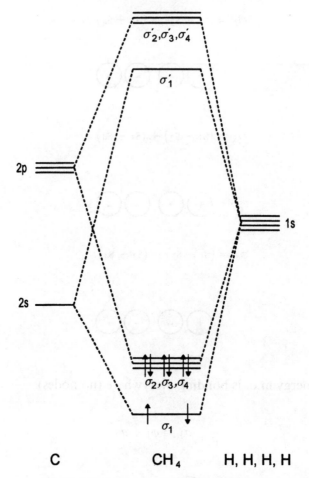

Fig. 7.12 Valence m.o. diagram of CH_4.

7.3 Other molecules and quantitative m.o. theory

The ethane molecule, C_2H_6

The ethane molecule still possesses sufficient symmetry to be dealt with in a manner similar to the previous examples. In particular, C_2H_6 can be thought of as constructed from two methyl groups for which the general form of the molecular orbitals is similar to that of NH_3. Accordingly, the bonding and anti-bonding m.o.s of ethane can, to a first approximation, be constructed from in-phase and out-of-phase combination of the σ_1, σ_2, σ_3 and σ_{nb} occupied m.o.s of two CH_3 groups similar to those of Fig. 7.11. This is shown in Fig. 7.13.

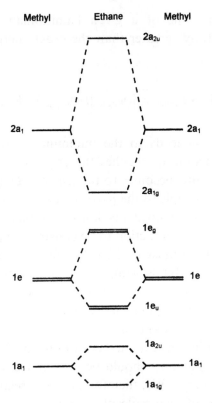

Fig. 7.13 Valence m.o.s of ethane constructed from m.o.s of two methyl groups.

Depending on the relative orientation of the two methyl groups, the total electronic energy is different. This means that rotation about the carbon–carbon axis is not 'free', but involves an energy barrier. The maximum occurs for the eclipsed conformation. For a simple m.o. treatment of the barrier to internal rotation in the ethane molecule, see ref. 80.

The variation principle

Whenever the molecule does not have sufficient symmetry, the determination of the atomic coefficients in the expressions of the various m.o.s

$$\psi_k = \sum_1^M c_{nk}\varphi_n \quad (k = 1, 2, \ldots, M) \ (n = 1, 2, \ldots, M) \qquad (7.24)$$

must be sought numerically. This is done by application of the *variation principle*, which was already mentioned in Chapter 5. This principle states that the expectation value of a given hamiltonian for an approximate wavefunction Ψ is always greater than the exact energy E associated with that hamiltonian:

$$[\langle\Psi|\hat{H}|\Psi\rangle/\langle\Psi|\Psi\rangle] > E_{\text{exact}} = [\langle\Psi_{\text{exact}}|\hat{H}|\Psi_{\text{exact}}\rangle/\langle\Psi_{\text{exact}}|\Psi_{\text{exact}}\rangle]. \qquad (7.25)$$

The exact wavefunction leads to the minimum energy and the more an approximate wavefunction approaches the exact one the closer the corresponding expectation value becomes to the true energy (for a recent application of the variation principle to the particle in a box problem, see ref. 81).

Therefore, the m.o.s obtained (as linear combinations of a.o.s) which enable an approximate wavefunction to be constructed (a Slater determinant for a closed-shell system) must lead to the minimum energy attainable for the basis set of atomic orbitals used, within the constraints of the method being applied. The larger the basis set, that is, the more flexible the wavefunction, the lower the calculated energy and, by implication, the better the wavefunction. Thus, m.o.s based on an infinite set of a.o.s would lead to the best wavefunction (within the eventual constraints of the hamiltonian used).

The coefficients in the m.o.s should be determined so as to minimize the molecular electronic energy. We, therefore, need to relate the total electronic energy with the energy of each molecular orbital.

The molecular electronic energy

As we have seen for polyatomic atoms in Chapter 5, the total electronic energy is not the sum of the energies of all electrons, because that would mean that the repulsion energy and the exchange energy would be counted twice. Thus, representing by ε_i the energy of occupied m.o. ψ_i (filled with two electrons in a closed-shell molecule) and by J_{ij} and $-K_{ij}$ the electron repulsion energy and the exchange energy for one electron in m.o. ψ_i and one electron in m.o. ψ_j, the total electronic energy E is

$$E = \sum_i^{\text{occ.}} \left[2\varepsilon_i - \sum_j^{\text{occ.}} (2J_{ij} - K_{ij}) \right]. \qquad (7.26)$$

It is noted that each electron in ψ_i suffers repulsion from the two electrons in ψ_j (hence the factor 2), but only its exchange with the electron of the same spin in ψ_j contributes to the energy.

The repulsion term J_{ij} is the classical Coulomb repulsion between the electron having a charge cloud given by $\psi_i^*\psi_i$ and each of the two electrons having charge cloud $\psi_j^*\psi_j$, and (apart from some constants) is given by the integral

$$J_{ij} = \langle\psi_i(1)\psi_j(2)|(1/r_{12})|\psi_i(1)\psi_j(2)\rangle. \tag{7.27}$$

The exchange integral, which arises from the antisymmetric nature of the wavefunction, differs by one permutation of the indices 1 and 2 for the electrons:

$$K_{ij} = \langle\psi_i(1)\psi_j(2)|(1/r_{12})|\psi_j(1)\psi_i(2)\rangle. \tag{7.28}$$

The energy ε_i of each m.o. is the sum of a term ε_i^0 representing the energy an electron in ψ_i would have if it was the only electron in the molecule (kinetic energy plus nuclear attraction terms) with the repulsion and exchange energies:

$$\varepsilon_i = \varepsilon_i^0 + \sum_j^{occ.}(2J_{ij} - K_{ij}). \tag{7.29}$$

Substituting ε_i of Eq. (7.29) into Eq. (7.26), we obtain

$$E = \sum_i^{occ.}\left[2\varepsilon_i^0 + \sum_j^{occ.}(2J_{ij} - K_{ij})\right], \tag{7.30}$$

and substituting the term in brackets of Eq. (7.29) into Eq. (7.26) we can also write:

$$E = \sum_i^{occ.}[\varepsilon_i + \varepsilon_i^0]. \tag{7.31}$$

The determination of molecular orbitals

The best m.o.s should lead to a minimum value of E (given by Eq. (7.31)). In the, so-called, *ab initio* (from the beginning) methods, in which the hamiltonian is explicitly written down and all the relevant integrals are computed, orbital wavefunctions are defined to minimize the full expression (7.31), with the ε_i^0 values computed from the m.o.s ψ_i. In many of the methods, minimizing

$$E = \sum_{i}^{\text{occ.}} \varepsilon_i \qquad (7.32)$$

is a common approximation made.

In the crudest methods, no hamiltonian operator is ever written out explicitly and no integrations are performed; the appropriate integrals are, in the end, taken as empirical parameters which are fitted from experimental data. The set of m.o.s is determined subject to the condition

$$\partial \varepsilon / \partial c_j = 0. \qquad (7.33)$$

Considering $\varepsilon = \langle \psi | \hat{H} | \psi \rangle / \langle \psi | \psi \rangle$ and substituting

$$\psi = \sum_{1}^{M} c_n \varphi_n \qquad (7.34)$$

the following expression is obtained:

$$\varepsilon = \frac{\langle \sum_n c_n \varphi_n | \hat{H} | \sum_n c_n \varphi_n \rangle}{\langle \sum_n c_n \varphi_n | \sum_n c_n \varphi_n \rangle} = \frac{\sum_m \sum_n c_m c_n \langle \varphi_m | \hat{H} | \varphi_n \rangle}{\sum_m \sum_n c_m c_n \langle \varphi_m | \varphi_n \rangle} = \frac{\sum_{mn} c_m c_n H_{mn}}{\sum_{mn} c_m c_n S_{mn}}$$

$$(7.35)$$

where the various interaction integrals have been abbreviated by H_{mn} and S_{mn} (overlap integral). The above expression can be written as

$$\sum_{mn} c_m c_n (H_{mn} - \varepsilon S_{mn}) = 0 \qquad (7.36)$$

which, after taking the derivative with respect to c_n keeping c_m constant, leads to

$$2 \sum_m c_m (H_{mn} - \varepsilon S_{mn}) - \left(\sum_m \sum_n c_m c_n S_{mn} \right) d\varepsilon / dc_n = 0 \qquad (7.37)$$

(a factor of 2 is included because c_m appears twice in the double summation, both as $c_n c_m$ and as $c_m c_n$). The condition (7.33) enables us to write

$$\sum_m c_m (H_{mn} - \varepsilon S_{mn}) = 0 \qquad (7.38)$$

for each coefficient c_n. There will, thus, be M simultaneous equations for M unknowns, which are called the *secular equations* (a name imported from celestial mechanics).

There is an alternative way of obtaining the secular equations without derivation. Let us substitute Eq. (7.34) in the wave equation $(\hat{H} - E)\psi = 0$:

$$\sum_n c_n(\hat{H} - E)\varphi_n = 0. \tag{7.39}$$

Multiplication by φ_m from the left

$$\sum_n c_n\varphi_m(\hat{H} - E)\varphi_n = \sum_n c_n[(\varphi_m\hat{H}\varphi_n) - E\varphi_m\varphi_n] = 0 \tag{7.40}$$

and integration gives

$$\sum_n c_n(H_{mn} - ES_{mn}) = 0. \tag{7.41}$$

The values of E above must satisfy the wave equation, which, in general, requires that the basis set in Eq. (7.34) is infinite.

A trivial solution of the secular equations is that all the unknowns c_n vanish. Any other physically meaningful solution requires that the determinant of the coefficients of the unknowns vanishes. That is,

$$|H_{mn} - \varepsilon S_{mn}| = 0. \tag{7.42}$$

This is called the *secular determinant*.

This $M \times M$ determinant is expanded into a polynomial of power M, and solving Eq. (7.42) leads to M values for the energy (one for each m.o.) providing the integrals H_{mn} and S_{mn} are known. By substituting each value of ε_i into Eq. (7.38) and introducing the normalization condition for each m.o.

$$\sum_{mn} c_m c_n \langle \varphi_m | \varphi_n \rangle = 1 \tag{7.43}$$

the several coefficients c_n are calculated for each ε_i value.

In Chapter 9, we will apply this methodology to the study of conjugated π systems. By adopting several approximations and using approximate parameters for the integrals H_{mn} and S_{mn}, molecular orbitals are easily calculated (Hückel method).

Molecular orbitals at different levels of approximation

It is in the definition of the integrals H_{mn}, or the lack of it, that most m.o. methods differ. The overlap integrals are easy to compute and were, indeed, tabulated for Slater-type atomic orbitals a long time ago (ref. 79). In some simple calculations, namely *extended Hückel-type calculations* a rough estimation of H_{mn} is as follows (see also page 79).

(a) H_{mm} – the *Hückel Coulomb integral* – is taken to be equal to the energy of an electron in the φ_m a.o. of the isolated atom set as minus the value of the corresponding ionization potential (Koopmans' theorem), or the configurationally averaged ionization potential.

(b) H_{mn} – the *Hückel resonance integral* – is taken as linearly dependent on the overlap integral S_{mn} and on the arithmetic or geometric mean of the energies corresponding to the φ_m and φ_n a.o.s of the appropriate atoms:

$$H_{mn} = kS_{mn}(H_{mm} + H_{nn})/2$$
$$H_{mn} = kS_{mn}(H_{mm}H_{nn})^{1/2}. \qquad (7.44)$$

The scaling factor is usually put equal to 1.75. Since H_{mm} and H_{nn} are negative quantities, H_{mn} is negative if S_{mn} is positive.

This method was systematically developed and applied by the American chemist Roald Hoffmann (1937– , 1981 Nobel laureate in Chemistry) (ref. 82).

Those methods which explicitly write out the hamiltonian, and aim at calculating the H_{mn} integrals (*ab initio* methods) frequently approximate the Slater-type a.o.s used as the basis set in SCF calculations (either as a minimal basis if only one function is used for each a.o., or a multiple set) by fitting gaussian functions to them. In this manner, the variable r no longer appears as an exponent and the integrals become easier to compute. For example, a STO-4G calculation uses four gaussians to approximately fit a minimal basis set of Slater-type orbitals.

When due account is taken of the form of the hamiltonian, the operator $1/r_{12}$ leads to polycentric integrals over a.o. functions which are directly related to J_{ij} and K_{ij}, for example

$$\langle \varphi_m(1)\varphi_u(2)|(1/r_{12})|\varphi_n(1)\varphi_v(2)\rangle. \qquad (7.45)$$

These integrals are particularly difficult to evaluate if the atomic functions are centred on two or more different atoms. Various approximations have been introduced in a number of variants, especially by the British chemist John A. Pople (1925– , 1998 Nobel laureate in Chemistry) and co-workers, and by the American chemist Michael J.S. Dewar based on the idea of *neglect of differential overlap* (ref. 83). In these methods an attempt is made to keep close to *ab initio* calculations. In some cases, however, parameters are chosen to give the best fit of experimental data (for example, enthalpies of formation) and we are then in the realm of *semi-empirical methods,* the grandfather of which is the Hückel method (see Chapter 9).

When running an *ab initio* or close to *ab initio* calculation, a computer program begins by determining the integrals required in building up H_{mn} and S_{mn} using estimated trial coefficients. Then the secular determinant is diagonalized to produce a set of orbital energies and first-improved coefficients. As mentioned previously when dealing with Hartree–Fock SCF calculations in atoms, this process is repeated until self-consistency is achieved.

For a review and thorough comparison of the various *ab initio* and semi-empirical methods, see ref. 84.

Nowadays, user-friendly software products are easily available, under the general name of molecular modelling packages, that enable any chemist to run powerful calculations without having to be a theoretician or a computational chemist. In addition, they frequently include very useful graphic features. Some of these packages can be taken freely through the Internet, such as the *ab initio* package GAMESS (General Atomic and Molecular Electronic Structure System) and the visualization package MOLDEN, for displaying MOLecular DENsity, which we have used to construct the plots of molecular orbitals in Chapters 6 and 7. A simple basis was used throughout: HF/3-21G. Figure 7.14 shows an outer isosurface for the highest occupied molecular orbital of ethanol calculated using the same method.

Pictorial approaches to molecular structure and reactivity based on computer-generated models of electron density distributions (as well as the

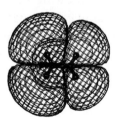 

CH₃ – CH₂OH

Fig. 7.14 Representation of the highest occupied m.o. (HOMO) of ethanol by an outer isosurface, for two orientations of the molecular skeleton.

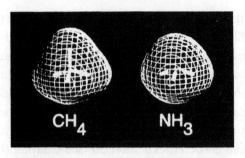

Fig. 7.15 Isodensity models for CH_4 and NH_3 (reprinted with permission from ref. 86, *Journal of Chemical Education*, **74**, 771 (1997); copyright 1997, Division of Chemical Education Inc.).

associated electrostatic potentials) are valuable for both experienced chemists and students, provided a critical use is made of those models. For a collection of more than 1500 photographs of molecular orbital representations, see ref. 85. The use of electron density models in the teaching of chemistry was recently addressed in ref. 86, by making use of a powerful molecular modelling package which is commercially available under the name SPARTAN (Fig. 7.15). More recently, and updating an early work by W. J. Jorgensen and L. Salem already quoted (ref. 69), a book with a CD-rom was published including a large number of examples manipulated as VRML (virtual reality modelling language) objects on the CD (ref. 162).

Electron correlation

As mentioned before, even the best Hartree–Fock SCF result for a molecule (or atom) is still in error, mainly due to the neglect of electron correlation. One way of improving the wavefunction is by allowing the interaction of configurations of the same symmetry. Formally, the process consists of applying the variation principle to determine the coefficients in a linear mixing of a ground state wavefunction Ψ_0 with appropriate excited configurations Ψ_i of the same symmetry:

$$\Psi_{CI} = a\Psi_0 + b\Psi_1 + c\Psi_2 + \cdots \tag{7.46}$$

On page 111 we made a brief reference to another way of dealing with electron correlation in atoms and molecules: density-functional theory.

In this theory, the analogue of the time-independent Schrödinger equation is

$$v(\mathbf{r}) + \delta F[\rho]/\delta\rho(\mathbf{r}) = [\delta E/\delta\rho]_{v(\mathbf{r})} \tag{7.47}$$

where the second member is the electronic chemical potential (6.21), $v(\mathbf{r})$ is the external potential (i.e. due to the nuclei) and $F[\rho]$ is the Hohenberg–Kohn functional of the electron density $\rho(\mathbf{r})$. For a recent account of density-functional studies of molecular systems, see ref. 87.

Molecular geometry

If we want to calculate a molecular geometry or reproduce the experimentally determined data, m.o. calculations should in principle be run for different arrangements of the nuclei in order to obtain the minimum energy geometry. Figure 7.16 shows the results of SCF *ab initio* calculations for H_2O and H_2S, as a function of the bond angle, keeping the bond lengths at the experimental value. The minima occur, respectively, for 104° and 92°, in good agreement with the experimental values 104.5° and 92.2°. For bigger molecules the situation is much more complicated and the so-called molecular mechanics methods are the most appropriate to use (see Section 8.5).

Since there are plenty of experimental data, molecular geometries are more often used to confirm the validity of calculations than calculations are used to determine geometries, with an important exception: conformational analysis. As an example of this field, Fig. 7.17 shows the results of a

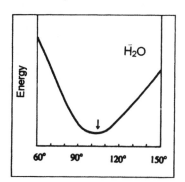 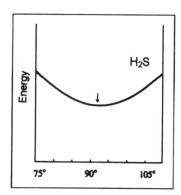

Fig. 7.16 Energy of H_2O and H_2S as a function of the bond angle, for bond lengths equal to the experimental values. (Redrawn with permission from ref. 88.)

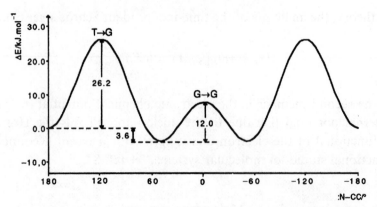

Fig. 7.17 Potential energy profile for the internal rotation of dimethylethylamine around the N–CC bond axis (Reprinted from ref. 89 with permission from Elsevier Science.)

Hartree–Fock SCF calculation using a gaussian basis set (HF/3-21G) of the potential energy curve associated with the internal rotation of dimethylethylamine.

8

Molecular orbitals and electron pair bonding

8.1 Atoms in molecules and structural formulae

When atoms combine with each other to form molecules, they partly loose their individuality but retain major characteristics, particularly those which are a direct consequence of the fact that atoms possess a nuclear structure. For example, atomic mass is not altered upon combination, an assumption advanced by Dalton well before Rutherford's alpha-scattering experiments and his proposal of the nuclear atom. The same is of course true of nuclear charge. A shell structure for the electrons of atoms leads to the considera-tion of core and valence electrons, the former being almost unaffected by atom association. It is because of the constancy of the characteristics just mentioned that we can refer, for example, to the element oxygen in the water molecule. Structural formulae keep track of this by using the symbols of the elements. The changes occurring at the valence electronic shells lead-ing to the existence of a network of bonds are symbolically represented by drawing strokes – one for each pair of valence electrons – between some of those symbols according to the geometry of the molecule. Thus, the four equivalent CH bonds in the methane molecule involve the eight valence electrons in four pairs:

$$
\begin{array}{c}
\text{H} \\
| \\
\text{H} - \!\!\!\!\overset{\displaystyle \text{C}}{\underset{\displaystyle |}{}}\!\!\!\! - \text{H} \\
\text{H}
\end{array}
$$

As there is no bonding interaction between the (peripheral) H atoms, no stroke is shown connecting their symbols. A similar representation is given for the isoelectronic and isogeometric species NH_4^+. The removal of a hydrogen ion, H^+, from the ammonium ion yields NH_3 whose structural formula now includes a stroke located on the N symbol corresponding to one electron pair of essentially non-bonding character out of the total of eight valence electrons:

$$H-\overset{\displaystyle \overline{N}}{\underset{\displaystyle H}{|}}-H$$

Single bonds are also considered in the water molecule, which means two non-bonding electron pairs:

$$H-\overset{\displaystyle \hat{O}}{\diagdown}-H$$

Although the use of strokes to represent bonds between atoms in molecules comes from the nineteenth century, the electron pair concept as necessary for the understanding of chemical bonding was introduced by G.N. Lewis (1875–1946) in 1916 (ref. 90) following Bohr's, then recently proposed, model of the atom. Indeed, the Lewis model still lies at the basis of much of present-day chemical thinking, although it was advanced before both the development of quantum mechanics and the introduction of the concept of electron spin. In a more quantitative way, it found a natural theoretical extension in the valence-bond approximation to the molecular wavefunction, as expressed in terms of the overlap of (pure or hybridized) atomic orbitals to describe the pairing of electrons, coupled with the concept of electron spin.

Structural formulae, by implying and representing a network of bonds, in accordance with the basic molecular geometry, play an extremely important role in the whole of chemistry. In particular, an identical representation of a given pair of atoms in various molecules means similar properties of such a group – bond length, bond energy, bond force constant, etc. – and enables an interpretation of additivity schemes based on the transfer of atoms and their properties between molecules. Comparable representations facilitate interpretation of differences; for example, the different polarities of CH, NH and OH (single) bonds, respectively, in CH_4, NH_3 and H_2O, can be related mainly to the differences in nuclear charge on going from C to N and O which are reflected in different electronegativities.

The success of such a well-established representation is obtained in spite of the fact that structural formulae do not even allow a proper visualization of the actual (van der Waals) shapes of the molecules. These are represented by envelopes of constant electronic charge density averaged over thermal motions of the nuclei, as given experimentally by X-ray and electron diffraction measurements or determined theoretically by molecular orbital or density-functional calculations. In particular, the topological properties of the total electron density offer no direct evidence to support a model that assumes the existence of discrete localized pairs of valence electrons of opposite spin, either bonding or non-bonding. In addition, the molecular orbital (m.o.) approximation to the wavefunction of a polyatomic molecule involves linear combinations of all atomic orbitals (a.o.s) which leads to non-localized molecular orbitals (with the exception of non-bonding m.o.s arising from symmetry considerations), in the sense that they do not just involve each atom and a neighbour. However, *neither the realistic models now available of molecules based on charge density nor the increasing ease of sophisticated m.o. calculations can replace the use of structural formulae in chemical thinking.* A bridge between the properties of electronic charge density and structural formulae-related models has been thoroughly established by Richard F.W. Bader (1931–) which provides a physical basis for the success of the latter (ref. 91). In Section 8.5 we will explore the mathematical basis from the point of view of molecular orbital theory.

8.2 The theory of atoms in molecules

In Bader's work, atoms in molecules are objects in real space which are defined through a unique partitioning of the molecule as determined by the topological properties of the molecular charge distribution. Like free atoms, a bound atom is defined as the union of an attractor (nucleus) and its associated basin. *The boundaries of this basin across which the atom, as an open quantum subsystem, can exchange charge and momentum with its neighbours, are not arbitrary; rather, they must be defined so that the expectation values of observables, quantum mechanically obtained for each subsystem, are of physical significance.* This means, for example, that the virial theorem applies to each atom as to the total system. We have then the only non-arbitrary way of partitioning potential energies of interaction between subsystems. The properties of such topologically defined atoms will be directly determined by their corresponding charge distribution. In particular, a constancy in the distribution of charge for an atom in a set of molecules leads to a corresponding constancy in the average electronic kinetic energy of the atom.

According to the virial theorem, the total (kinetic + potential) energy for a system with inverse square forces is equal to minus the average kinetic energy, which means that the average potential energy is minus twice the average kinetic energy. Therefore, if the distribution of charge for an atom is identical in two different molecules, the atom – possessing a constant average kinetic energy – will contribute identical amounts to the total energy in both molecules.

An important conclusion of an atomic statement of the virial theorem is that the total energy of a molecule is expressible as a sum of atomic energies. In more general terms, each topologically defined atom makes an additive contribution to the average value of every molecular property.

The electron density ρ of a molecule is a physical quantity which has a definite value $\rho(\mathbf{r})$ at each point of coordinates $\mathbf{r}$ in three-dimensional space. The topological properties of this electronic charge distribution can be summarized in terms of its critical points: maxima, minima and saddles. Figure 8.1 displays the electronic charge density in three planes of the ethylene molecule.

As expected, the electron density has maxima at the positions of the nuclei. As a matter of fact, the electron density distribution for any molecule contains the information needed to determine the distribution of nuclear charge as well. It is found that the electron density at the nuclei $\rho(0)$ is related to the atomic number Z through the empirical expression $\rho(0) = 0.4798 Z^{3.1027}$ atomic units, with a mean deviation of 2.6%, for $Z < 55$ (ref. 89).

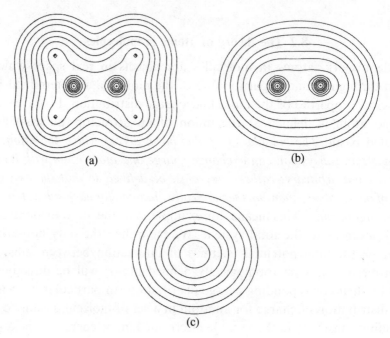

Fig. 8.1 Contour plots of the electronic charge density in different planes of the ethylene molecule: (a) the plane of the nuclei; (b) the plane obtained by a 90° rotation about the C–C axis of the plane shown in (a); (c) the plane perpendicular to the C–C axis at its midpoint. The values of the contours increase from the outer contours inwards. (Adapted with permission from Bader *et al.* ref. 92; copyright 1981, Institute of Physics Publishing.)

Returning to Fig. 8.1, special attention is drawn to the midpoint $\mathbf{r_c}$ in the CC axis. The corresponding value ρ is a minimum or a maximum depending on the axis considered. The first derivatives of ρ vanish in any case, but the second derivatives are positive for the CC internuclear axis (positive curvature) and negative for the two other cartesian axes. This shows the need to know not only the gradient vector field associated with ρ

$$\nabla\rho = \mathbf{i}\,\partial\rho/\partial x + \mathbf{j}\,\partial\rho/\partial y + \mathbf{k}\,\partial\rho/\partial z \tag{8.1}$$

but also the laplacian of ρ

$$\nabla^2\rho = \nabla\cdot\nabla\rho = \partial^2\rho/\partial x^2 + \partial^2\rho/\partial y^2 + \partial^2\rho/\partial z^2. \tag{8.2}$$

Similar critical points occur between the carbon and hydrogen nuclei but not necessarily at the midpoint in the CH axis. These points belong to the interatomic surfaces which define the boundaries of the basin of each atom. The whole surface $S(\mathbf{r})$ of an atom is such that the vector $\nabla\rho(\mathbf{r})$ is tangent to it at every point $\mathbf{r}$. In Bader's theory, this is called a zero-flux surface in the gradient vectors of the charge density. Figure 8.2 shows the intersections of the interatomic surfaces with the plane of the nuclei of C_2H_4 superposed on a contour map of the electron density; it also shows the five 'bond paths' defined as the lines of maximum charge density linking the nuclei.

It is noted that, contrary to isolated atoms, the zero-flux surfaces of bound atoms are in general not spherical.

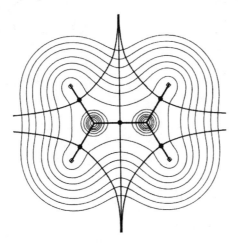

Fig. 8.2 Intersections of the interatomic surfaces with the plane of the nuclei of the ethylene molecule superposed on a contour map of the electron density. Also shown, the five 'bond paths' (adapted with permission from ref. 92; copyright 1981, Institute of Physics Publishing).

It is the use of zero-flux surfaces for the topological definition of atoms or functional grouping of atoms in molecules that maximizes the extent of transferability of their properties between systems.

Figure 8.3 shows the contour plots of $\rho(\mathbf{r})$ for the nuclear planes of CO, H_2O, H_2CO, as well as the bond paths and the interatomic curves as intersections of the interatomic surfaces with that plane. Similar representations are given for pentane and hexane in a conformation where all C nuclei and a terminal H nucleus of each methyl group are in the plane of the diagram.

It is found that multiple bonds do not appear as such in the topology of the electron density. However, the value of the charge density at a bond critical point reflects the bond multiplicity and can indeed be empirically correlated with bond orders (refs. 93 and 94). As expected, it is also found that the charge distribution in the CC interatomic surface in ethylene has an elliptical nature associated with the presence of a π bond.

The relative abilities of two bonded atoms to attract and bind electron density within their basins are a measure of their relative electronegativities. It is found, for example, that H is slightly more electronegative than C in saturated hydrocarbons and that the electronegativity of C relative to H increases with the degree of unsaturation and with the extent of geometric strain.

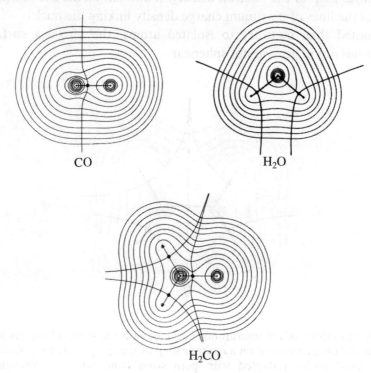

CO H_2O

H_2CO

Fig. 8.3 (*see opposite for caption*)

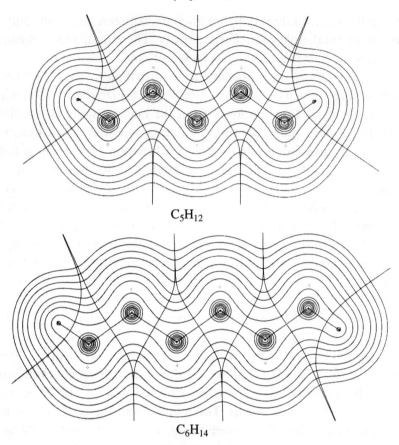

C_5H_{12}

C_6H_{14}

Fig. 8.3 Contour plots of $\rho(\mathbf{r})$ for the nuclear planes of CO, H_2O, H_2CO, and for C_5H_{12} and C_6H_{14} in a conformation where all C nuclei and a terminal H nucleus of each methyl group are in the plane of the diagram. The bond paths and the interatomic curves as intersections of the interatomic surfaces with the plane of the nuclei are also shown. The values of the contours increase from the outer contours inwards (adapted with permission from ref. 91; copyright 1995, Oxford University Press, and ref. 92; copyright 1981, Institute of Physics Publishing).

It is also found that a significant degree of interatomic charge transfer is accompanied by a polarization of the valence electron densities of the atoms in the opposite direction, due to the electric field created by the charge transfer. This affects the measured dipole moments. For example, CO has a near-zero dipole moment because of the very significant polarizations of the atomic densities, particularly that of the carbon atom, which oppose the pronounced charge transfer moment.

In chemistry, the energy changes of interest are but small fractions of the total energy of the molecules. Similarly, as pointed out by Bader, the changes induced in the electron density by interactions between atoms are only small 'ripples' in the total molecular charge density. These small changes can however be made more evident by the laplacian of $\rho(\mathbf{r})$.

For isolated atoms, it is the laplacian distribution that displays shell structure in three-dimensional space, unlike the distribution of $\rho(\mathbf{r})$ which is a monotonous decay function away from the nucleus along any radial line; a radial distribution plot also shows extrema but it is a one-dimensional function. A negative second derivative of a function $f(x)$ means a negative curvature of $f(x)$, with $f(x)$ at x being greater than the average of its values at the neighbouring points $x + \mathrm{d}x$ and $x - \mathrm{d}x$; it is said that $f(x)$ is concentrated in regions where $\mathrm{d}^2 f(x)/f(x)\,\mathrm{d}x^2 < 0$, even though this does not imply the existence of a maximum in the function itself (may be a shoulder). Analogously, *there is charge concentration in the regions of free or bound atoms where $\nabla^2 \rho(\mathbf{r}) < 0$, and charge depletion where $\nabla^2 \rho(\mathbf{r}) > 0$, irrespective of the occurrence of extrema in the charge density.*

A local maximum in $-\nabla^2 \rho(\mathbf{r})$ is a maximum in the concentration of electronic charge. For example, in CH_4 there are four tetrahedrally arranged concentrations of charge (in the sense above) together with centres of local charge depletion in each of the triangular faces of the tetrahedron. Similarly, the oxygen atom of the H_2O molecule exhibits four local concentrations of valence shell charge: two bonded concentrations and two larger and equivalent non-bonded concentrations. The number and the relative positions of the local maxima in the valence-shell charge concentration of a central atom in a molecule are a consequence of the partial localization of electron pairs in its valence region which results from the field of the neighbouring atoms operating in concert with the Pauli principle.

Problem 8.1   Describe the number, relative positions and sizes of local concentrations in the molecule NH_3.

Although there is no correspondence between these local charge concentrations and the electron pairs of the Lewis model, there is a clear relation between the number of local charge concentrations and an electron count. Also, for example, it is remarkable that a Lewis base or a nucleophile implies a local charge concentration, while a local charge depletion is a Lewis acid or electrophile. Additional intellectual satisfaction is obtained from the demonstration that the total energy of a AX_n molecule is minimized for the geometry which maximizes the separations between the charge concentrations for the central atom.

For further reading on the theory of atoms in molecules see ref. 91.

8.3 Structural formulae and non-independent bonds

All the information that can be known about a system in a given stationary state is contained in the state function $\Psi(\mathbf{r})$, particularly molecular electron densities. The use of electron density to interpret atoms in molecules, bonds and structures constitutes a bridge between the concept of state function and the physical model of matter in real space.

It is found that, in general, the molecular charge densities obtained from single determinant SCF calculations carried essentially to the Hartree–Fock limit (see page 163) are accurate to within 2–5% of the total charge density at any point in space, except perhaps for very large distances from the nuclei. Coulomb electron correlation effects are relatively small; in particular, the value of $\rho(\mathbf{r})$ in the neighbourhood of a bond critical point calculated in that way exceeds the value obtained from a correlated wavefunction by only a few per cent.

Whatever the level of approximation used in the calculation, in general the molecular orbitals (m.o.s) for polyatomic molecules have contributions from the orbital functions of several atoms. They are non-localized, in the sense that they do not just involve each atom and a neighbour, as might be suggested by the traditional structural formulae. An exception is provided by the occurrence of localized m.o.s as non-bonding orbitals, namely for symmetry considerations, as is the case for the non-bonding electron pair of π symmetry in the H_2O molecule studied in the previous chapter.

Although in some cases there is a close relation between the concentration of charge predicted by the laplacian of $\rho(\mathbf{r})$ given by Eq. (8.2) and the symmetry properties of the highest occupied and lowest unoccupied m.o. (HOMO and LUMO), one of the virtues of the orbital model is that the relative phases of the orbitals determine their symmetry properties and their effective overlap (as we have already seen in Chapter 7 and will return to in future chapters).

The relation between delocalized m.o.s and structural formulae can begin by just counting the total number of bonding and anti-bonding electrons, N_b and N_{ab}, respectively, and then determine what could be called a 'global bond order' as the sum of the multiplicities of the several bonds in the molecule:

$$\text{global bond order} = (N_b - N_{ab})/2. \tag{8.3}$$

By using molecular symmetry considerations and some well established patterns (e.g. the general occurrence of single bonds to H atoms), individual bond multiplicities can be interpreted. Thus, for CH_4 with its four equivalent bonds, eight electrons are associated with four valence non-localized bonding m.o.s, which implies a mean effect of two bonding electrons per CH section: four single bonds. For ethylene, with all 12 valence electrons as bonding electrons and two types of bonds, the four (single) CH bonds correspond to the global effect of four × two electrons, leaving four for the CC bond, a double bond:

$$
\begin{array}{c}
H H \\
\diagdown \diagup \\
C{=}C \\
\diagup \diagdown \\
H H
\end{array}
$$

In the benzene molecule there are 30 valence bonding electrons, the global effect of 12 of them being assigned to the six C–H bonds which leaves a global effect of 18 electrons for the six equivalent CC bonds, i.e. a bond multiplicity of 1.5, the mean value for single and double bonds:

As a rule, all valence electrons in hydrocarbons (in their ground states) are assigned to bonding m.o.s as a result of various features: an even number of valence m.o.s, half of them bonding and half anti-bonding, the energies of the former all lying below the anti-bonding ones, and a number of valence electrons (equal to the number of contributing a.o.s) that completely fill the bonding m.o.s.

Problem 8.2   Interpret the structural formula of propene applying the model above to the chemical formula C_3H_6 and knowing that one CC bond is appreciably longer than the other.

Cyclobutadiene, C_4H_4, and cyclooctatetraene, C_8H_8, have been cited as exceptions to the rule, both with two non-bonding π molecular orbitals if regular geometries are assumed (see page 227). The former, first synthesized in 1965, is a highly reactive hydrocarbon that dimerizes very easily. Experimental observations and theoretical calculations are now in agreement that cyclobutadiene has a rectangular geometry with CC bond lengths close to the typical single- and double-bond values. Similarly, cyclooctatetraene has a non-planar geometry with alternating single and double bonds.

In any case, it must be kept in mind that bond multiplicities, or bond orders in current terms, do not usually coincide with bond orders of molecular orbital theory calculated from the a.o. coefficients in the expressions of the occupied m.o.s. This point will be resumed in Chapter 9 in the context of π bonding.

The occurrence of occupied non-bonding molecular orbitals is a frequent feature of molecules possessing atoms with more valence electrons than valence a.o.s. Thus, for NH_3, for example, with an odd number of m.o.s, the total bond multiplicity is 6, corresponding to six bonding electrons out of a total of eight. For the isoelectronic molecule H_2O, although the total number of valence m.o.s is even, there are two (filled) non-bonding m.o.s and only four of the valence electrons assigned to bonding m.o.s, hence the structural formula

For more complicated molecules, this simple model based on electron count still holds, relating molecular orbitals and structural formulae. However, the decision on which types of m.o.s are occupied may not be as straightforward as for hydrocarbons and the simple molecules above.

This link between molecular orbital theory and structural formulae is based on the total net bonding effect of the valence electrons, just as for diatomic species like N_2, O_2, F_2, or O_2^{2-}

$$|N\equiv N| \qquad \overline{O}{=}\overline{O} \qquad |\overline{F}{-}\overline{F}| \qquad |\overline{O}{-}\overline{O}|^{2-}$$

where bond features (e.g. bond energies, bond lengths and stretching force constants) can so be interpreted. The only difference, in the case of polyatomic molecules, is that the global net bonding effect is split among the various bonds; and this is carried out by taking into account experimental information about the bonds themselves. The structural formulae reflect similarities and differences in these bond features. Accordingly, such features as are reflected in the formulae can be transferred from molecule to molecule. For example, the C–H bond length in hydrocarbons is always around 108 pm, its stretching force constant around $500\,N\,m^{-1}$, and its bond energy about $405\,kJ\,mol^{-1}$. For these properties, each formal bond behaves independently. In particular, bond energies can be added to give the total energy of the molecule relative to the corresponding separate atoms.

However, inasmuch as the structural formulae are related to molecular orbitals in a way that has no bearing on individual m.o. energies (except in relative terms in order to decide which m.o.s are occupied in the ground state), the structural formulae do not give any information about electron energies and properties related to them. For example, just as the formula of F_2 cannot be taken as six filled non-bonding m.o.s of equal energy, the structural formula of methane does not entitle us to infer a valence-shell electronic configuration of CH_4 consisting of four equivalent bonding m.o.s, each one occupied by two paired electrons.

In fact, the photoelectron spectrum of the valence electrons for this molecule has two bands (2.0–2.6×10^{-18} J and 3.6–3.9×10^{-18} J) (Fig. 8.4) which shows that the eight valence electrons must be associated with two energy values. This is in qualitative agreement with the m.o. diagram we have established in the previous chapter (Fig. 7.12). Similarly, the classical structural formula for NH_3 does not allow the consideration of six bonding electrons with the same energy plus two non-bonding electrons with a different (higher) energy:

Indeed, the corresponding photoelectron spectrum indicates three different energy values for the eight valence electrons, in accordance with the m.o. diagram of Fig. 7.11. The same problem arises for H_2O. Its classical formula does not mean four isoenergetic bonding electrons and four isoenergetic

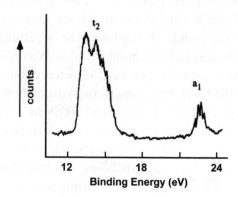

Fig. 8.4 Photoelectron spectrum of the valence electrons for methane (ref. 95).

non-bonding electrons (at higher energy):

As we have seen previously (Fig. 7.8), the corresponding photoelectron spectrum has four bands demonstrating the existence of four different energy values for the set of eight valence electrons, in agreement with the m.o. diagram of Fig. 7.7.

This means that *for some purposes, namely electronic configurations and related properties, the chemical bonds (and electron lone pairs of the same atom) in structural formulae cannot be regarded as independent from each other.*

Related to this, and as pointed out in ref. 96, is the fact that the molecules CH_4 and H_2O do not possess just one stretching vibration for C–H or O–H, respectively, but two. In each molecule, the bonds are not independent. For example, it is not legitimate to consider two separate stretching OH vibrations in the water molecule as the O atom would be common to both. The interaction between the corresponding force fields leads to two different vibrational frequencies: one (ω_s) describes the in-phase vibrations (symmetric stretching)

whereas the other (ω_a) describes the out-of-phase vibrations (asymmetric stretching).

The symmetric and the asymmetric vibrations are the true independent stretching vibrations, and hence are those that lead to distinct absorption bands in the infrared spectrum (in addition to the independent bending vibration).

Similarly, if we begin by considering two electronic functions, f_1 and f_2, each one associated to each formal bond in H_2O, such functions will be equivalent but not independent, in the sense that an 'interaction' between them must be taken into account. Such an interaction will be characterized later in the chapter, but we can already anticipate that it leads to a symmetric function

$$\sigma_s = (f_1 + f_2)/N_s \tag{8.4}$$

and an asymmetric function

$$\sigma_a = (f_1 - f_2)/N_a \tag{8.5}$$

truly independent from each other and corresponding to different energies. N_s and N_a are normalization factors. The same reasoning applies to two functions, n_1 and n_2, chosen to describe equivalent but non-independent electron lone pairs:

$$\sigma_{ns} = (n_1 + n_2)/N_{ns} \tag{8.6}$$

$$\sigma_{na} = (n_1 - n_2)/N_{na}. \tag{8.7}$$

The occurrence of four bands in the photoelectron spectrum of H_2O, for the valence electrons, can thus be made compatible with the classical structure for the molecule, although a more rigorous approach should also consider possible interactions between functions f_i and n_i, as we shall see later in this chapter. We shall further recognize that, within the molecular orbital framework, the four independent functions are, in fact, the (non-localized) valence molecular orbitals.

8.4 Orbitals and electron pairing in valence-bond theory

The first electronic function to be defined in direct association with a formal chemical bond was, as expected, for the H_2 molecule. It marked the beginning of the so called *valence-bond theory*, in 1927, and was suggested by the Austrian physicist Walter H. Heitler (1904–1981) and by the German physicist Fritz W. London (1900–1964).

Firstly, it is recognized that, for two separate H atoms, A and B, with electrons 1 and 2, respectively, the electronic (wave) function can be expressed as the product of the respective 1s orbital functions

$$\varphi_{1sA}(1) \times \varphi_{1sB}(2). \tag{8.8}$$

Next, it is assumed that, when the atoms are associated to form H_2, this function would not need appreciable modification except for the possibility of assigning electron 1 to orbital 1s(B) and electron 2 to orbital 1s(A):

$$\varphi_{1sB}(1) \times \varphi_{1sA}(2). \tag{8.9}$$

This possibility arises because the electrons in the molecule are indistinguishable.

The wavefunction will then be comprised of a spatial part with bonding character, that is, describing increased electron density between the nuclei, with the following form

$$\psi_{vb} = \varphi_{1sA}(1)\,\varphi_{1sB}(2) + \varphi_{1sB}(1)\,\varphi_{1sA}(2) \tag{8.10}$$

(ignoring the normalization factor, for simplification). This function is symmetric regarding the exchange of the coordinates of the electrons. Since the Pauli principle requires all electronic wavefunctions to be antisymmetric, Eq. (8.10) must be multiplied by an antisymmetric spin-function, namely

$$\alpha(1)\beta(2) - \beta(1)\alpha(2). \tag{8.11}$$

That is, the two electrons must have opposite spins.

The spatial part (Eq. (8.10)) of the complete wavefunction can be compared with the corresponding part of the wavefunction expressed in terms of the bonding molecular orbital which is

$$\psi_1 = \varphi_{1sA} + \varphi_{1sB} \tag{8.12}$$

(again ignoring the normalization factor). The wavefunction expressed in terms of the occupied m.o.s ψ is given by the Slater determinant (see Chapter 6)

$$\Psi = \begin{vmatrix} \psi_1(1) & \psi_1(2) \\ \bar{\psi}_1(1) & \bar{\psi}_1(2) \end{vmatrix} = \psi_1(1)\bar{\psi}_1(2) - \bar{\psi}_1(1)\psi_1(2) \tag{8.13}$$

which, after substituting Eq. (8.12), becomes

$$\Psi = [\varphi_{1sA}(1)\,\varphi_{1sB}(2) + \varphi_{1sB}(1)\,\varphi_{1sA}(2) + \varphi_{1sA}(1)\,\varphi_{1sA}(2)$$
$$+ \varphi_{1sB}(1)\,\varphi_{1sB}(2)][\alpha(1)\beta(2) - \beta(1)\alpha(2)]. \tag{8.14}$$

The difference with respect to the spatial part of the valence-bond (v.b.) wavefunction lies in the two last terms in the first brackets. Whereas m.o. theory permits the two electrons to be associated with the same 1s function (A or B), this possibility is not considered by the v.b. expression (8.10). In this sense, it can be said that this form of valence-bond theory accounts (although exaggerating) for the Coulomb correlation of electrons according to which they tend to be as far apart as possible. Molecular orbital theory, in the form above, does not account for this correlation. One can expect that a situation between the two will in general occur.

The form of the v.b. wavefunction can be improved by 'mixing' contributions from functions describing ionic structures

$$H^+ \quad H^- \quad \text{and} \quad H^- \quad H^+$$

which assume the two electrons to be in the same 1s atomic orbital.

For more complex molecules, the v.b. wavefunctions are defined using functions describing each formal covalent bond which are similar to Eq. (8.10) (eventually mixed with ionic contributions). In this process, it is often necessary to consider, not the pure s and p (or s, p, and d) atomic orbitals, but linear combinations of them: *hybrid atomic orbitals*. We have already briefly encountered hybrid orbitals in the study of excited H atoms (for $n = 2$, for example) and in the m.o. study of BeH_2. For instance, in the case of CH_4, the 2s atomic orbital and the three 2p a.o.s are replaced by four linear combinations: *four sp^3 hybrid orbitals*. They are equivalent and have symmetry axes that conform to the actual tetrahedral geometry of the molecule, as is illustrated in Fig. 8.5.

Fig. 8.5 Carbon sp^3 hybrid orbitals (with their axes defining angles of 109° 27′) appropriate to the v.b. description of the four equivalent CH bonds of CH_4.

By overlapping each H 1s orbital with each one of these hybrid orbitals of C, in a manner similar to H_2, four equivalent functions are defined each localized in each formal CH bond: four *localized functions*, as graphically illustrated in Fig. 8.6. After eventual mixing of ionic contributions and the inclusion of the spin-functions, the global electronic wavefunction can be constructed.

The form of the wavefunction could be easily established if the four equivalent localized v.b. functions were independent. But the fact that there is not just one ionization energy for the eight valence electrons in CH_4 shows, once again, that these functions cannot be considered independent. This means, that, rigorously, we should take into account the interactions between them. In valence-bond theory, this is achieved by describing the

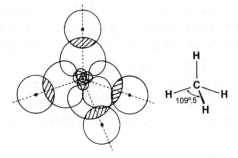

Fig. 8.6 Equivalent localized v.b. functions for the CH bonds of CH_4.

structure of CH_4 not only in terms of the *perfect pairing formula*

$$
\begin{array}{c}
H \\
| \\
H\!-\!\overset{\displaystyle C}{\underset{\displaystyle |}{}}\!-\!H \\
H
\end{array}
$$

but as a mixture (*resonance*) with less traditional formulae, such as

$$
\begin{array}{c}
H \\
\diagup \overset{\displaystyle C}{\underset{\displaystyle |}{}} \diagdown \\
H\quad|\quad H \\
H
\end{array}
$$

8.11

Thus, precisely as in m.o. theory, there will be no independent localized functions for each bond.

This process of constructing functions for the various resonant formulae, followed by an adequate combination of them, is mathematically more complex than the mathematics of molecular orbital theory. It is therefore understandable that, after the initial preference of chemists for the v.b. bond theory which has a closer relation to Lewis structures – especially due to the contribution of Linus Pauling – m.o. theory became increasingly popular. In addition, m.o. theory leads directly, not only to fundamental states (through the occupied m.o.), but also to excited states (through vacant m.o.) of molecules. In recent years, however, a new form of valence-bond theory has been developed that is more amenable to computation (*spin-coupled valence-bond theory*) in which the molecular wavefunction is expressed as a linear combination of all the coupling schemes of the various electrons corresponding to the same resultant spin (ref. 97).

As we have already stated, if it is true that some properties (in particular those related to the electron energies) can only be understood by considering non-independent bonds, in other cases (e.g. bond energies) such consideration is not important. For example, the C–C bond energies do not vary significantly from molecule to molecule, in agreement with a description in terms of the overlap of sp^3 atomic orbitals of adjacent C atoms, as illustrated next for ethane in a direct extension of the case of CH_4:

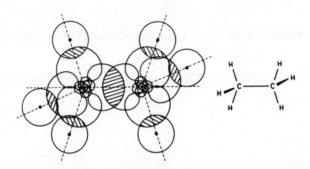

The total electron density of a molecule is also not very sensitive to the neglect of bond interaction. Hence, the polarity of molecules can be reasonably interpreted in terms of localized bond functions. For example, except for the polarization effects mentioned before (page 173), the molecular dipole moment of NH_3 can be interpreted in terms of the localized NH functions describing polar bonds plus the contribution of a non-bonding electron pair occupying a hybrid orbital of N:

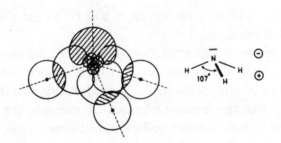

For the water molecule, we have, analogously

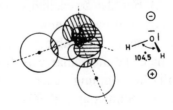

In these two cases, the hybrid orbitals of the central atom having axes that conform to the experimentally observed bond angles are not rigorously sp^3, but the differences are not important at this stage.

Other molecular geometries imply, in v.b. theory, other hybrid orbitals: sp^2 *hybrid orbitals*, three in number, having symmetry axes that make 120° angles,

or sp *hybrid orbitals*, two in number, having axes that make an angle of 180°:

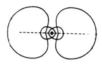

Examples are provided by the central atoms of NO_3^- and HCN, respectively.

Problem 8.3   What are the carbon hybrid orbitals appropriate, in v.b. theory, to the description of the structures of (a) CO_2? (b) CO_3^{2-}? (c) C_2H_4? (d) C_2H_2? (e) C_6H_6 (benzene)? (f) C_6H_{12} (cyclohexane)?

For a more detailed discussion of the interpretation of molecular geometry in terms of valence-bond theory at this level of approximation, see ref. 98.

8.5 Molecular geometry and the valence-shell electron pair repulsion model

The geometry of molecules can in principle be interpreted either in terms of forces or in terms of energy. We begin by recognizing that, for the equilibrium geometry of any molecule, the resultant of all electrostatic forces acting at the position of each nucleus, which includes internuclear repulsions and attractions between the nucleus and the whole electron charge distribution, must vanish. Should this not be the case then the nuclei would adjust themselves to another geometric configuration. According to what is known as the Feynman's electrostatic theorem, the force on a nucleus can be calculated from classical electrostatics if the charge distribution is known, even though this distribution is governed by quantum mechanics (ref. 99).

For example, for the H_2 molecule such a balance of forces occurs for an internuclear distance of 74 pm (the bond length of the molecule):

Nuclear repulsion

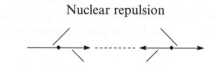

Resultant attraction of nuclei to the electrons

The geometric arrangement of the nuclei appropriate to the above balance of forces is also the geometry for which the molecular energy is minimum. The molecular energy considered here (although usually called the potential energy of the molecule) is the sum of the electronic (kinetic + potential) energy and the internuclear repulsion (potential) energy (see also page 71).

Thus, the understanding of molecular geometry requires a relationship to be drawn between a given geometric arrangement of the nuclei and the electronic spatial distribution and/or the molecular energy. The Walsh diagrams which we have previously considered for AH_2 molecules (Chapter 7) illustrate these relationships. Walsh diagrams are an example of qualitative and semiquantitative application of quantum mechanics to the interpretation of molecular geometry. They rely largely on the phase and symmetry properties of molecular orbitals. At a more quantitative level, the SCF m.o. calculations for the molecules H_2O and H_2S (Fig. 7.16) constitute basic examples. An accurate calculation of the energy of a polyatomic molecule requires not only calculations at the Hartree–Fock level involving a sufficiently large expansion of the wavefunction but also the inclusion of Coulomb correlation through configuration interaction. Alternatively, density-functional theory is being increasingly used. However, the equilibrium geometry found from a Hartree–Fock calculation without configuration interaction is often very similar to that found from the calculation when correlation is included. For molecules up to about 20 atoms, the most stable geometry can be computed with an accuracy comparable to experimental accuracy, in a reasonable length of time.

For large molecules, however, the computer requirements become increasingly prohibitive, especially when conformationally flexible compounds are tackled. Alternative approaches to quantum-mechanical methods are known, based on potential functions and parameters derived from detailed analysis of vibrational spectra. These so-called force field methods are now joined in what is called molecular mechanics, an empirical method that considers the molecule as a collection of spheres (possibly deformable) bound by harmonic forces (eventually corrected with cubic and quartic potentials). The energy of a molecule is calculated as a sum of the steric and non-bonded interactions for various possible geometries in comparison with an arbitrary reference geometry (refs. 100 and 101). This method is now used extensively and is supported by a large number of commercial software packages with user-friendly computer graphics. A critical analysis which identifies the inherent assumptions and the approximations in these methods can be found in ref. 102.

As far as qualitative quantum-mechanical models are concerned, the first attempts applied the principles of valence-bond theory. For example, the bent geometry of the water molecule was initially interpreted in terms of the overlap of the 1s atomic orbitals of the H atoms with two 2p atomic orbitals (with mutually perpendicular axes) of the O atom, each one contributing one electron:

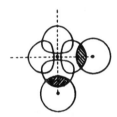

This description is supported by the fact that the (degenerate) states of lowest energy for an isolated O atom involve one 2p orbital fully occupied and two half-filled 2p orbitals.

The concept of hybridization of atomic orbitals was subsequently introduced, in an attempt to interpret the difference between the actual bond angle for the water molecule and the value of 90° considered in the previous model. This concept had already been introduced to interpret, for example, the tetrahedral geometry of the methane molecule. We shall come back to this subject later in the chapter, to conclude that, although it is possible to establish a *correlation* between molecular geometry and hybrid orbitals, it is not correct to take the latter as the basis of an *explanation* of the former. This distinction is very important in teaching.

The teaching of molecular geometry gains much from analogic considerations. These include the use of the Periodic Table of the elements: for example, H_2Te, H_2Se, H_2S are all bent molecules like H_2O; AsH_3 and PH_3 are molecules of pyramidal geometry like NH_3, etc. Another pattern is provided by the geometric relations in hydrocarbons associated with the 'coordination number' of each C atom, for example a linear arrangement in general when C is bonded to two other atoms.

However, the most successful and widely used model for the prediction of geometries of closed-shell molecules is the so-called 'valence-shell electron pair repulsion model' (VSEPR model) developed by R. Nyholm and R. Gillespie (ref. 103). In 1940, and as a natural extension of the localized electron pair model of Lewis, N. V. Sidgwick and H. M. Powell (ref. 104) suggested that the main contribution to the difference in energy between different geometric configurations of the nuclei of a molecule would arise from the repulsion of the valence-shell electron pairs. For example, CH_4 would be tetrahedral instead of planar, mainly because the tetrahedral geometry would correspond to a minimum repulsion of the four valence electron pairs. These ideas were later developed by Nyholm and Gillespie who added that the repulsions would be greater when electron lone pairs were involved. Thus, the decreasing order of repulsions would be:

lone pair−lone pair > lone pair−bonding pair > bonding pair−bonding pair.

In this way the decrease of the bond angles could be understood for CH_4 (109.5°), NH_3 (107.3°) and H_2O (104.5°). The planar geometry of XeF_4

$$\begin{matrix} F & & F \\ & \diagdown \overline{} \diagup & \\ & Xe & \\ & \diagup \overline{\overline{}} \diagdown & \\ F & & F \end{matrix}$$

would also be rationalized in terms of non-bonding lone pairs in an axial orientation which minimizes their repulsion energy.

Although the electron pair repulsions considered above are usually taken as electrostatic repulsions, Gillespie (refs. 105 and 106) has insisted that they should not be regarded as Coulombic interactions but as repulsions associated with spin correlation, hence, with the Pauli principle.

This latter interpretation follows a model developed by J.W. Linnett in 1964 (ref. 107) in which the orbital concept is largely ignored in favour of spin correlation which is a consequence of the antisymmetrization of the total wavefunction demanded by the Pauli principle. In such a model, what matters are the most likely relative positions of the electrons. It can be shown that, with an antisymmetric wavefunction, electrons having parallel spins tend to be as far apart as possible around the nucleus of an atom. Let us take the carbon atom as an example. For its excited valence configuration $2s^1$, $2p^3$, the four electrons have preferably parallel spins (extension of Hund's rule to excited configurations); and, among the infinity of spatial arrangements, the most likely ones are those in which the four electrons define the vertices of a tetrahedron centred at the nucleus. In particular, for a given distance between each electron and the nucleus, the arrangement

has a higher probability than, for example,

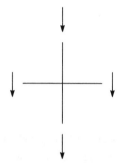

because the electrons have more separation in the first case. It is not surprising, therefore, that the association of four H atoms to a C atom leads preferably to a tetrahedral rather than a planar molecule.

Based upon Linnett's model, Gillespie begins by associating each unpaired electron with a region or 'domain' in the atom, and then defines 'electron pair domains' as regions of the atom in which there is a high probability of finding a pair of electrons of opposite spin, for example

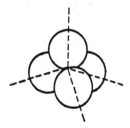

for O^{2-}, F^-, Ne. That is, the valence charge density is assumed to be spatially localized into pairs of electrons of opposite spin. The next step is to extend this concept of electron pair domains to bonding and non-bonding

electron pairs in molecules, for example (schemes reproduced by permission
of the Royal Society of Chemistry, ref. 105)

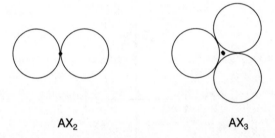

for CH_4, NH_3, H_2O (S = single bond, L = lone pair), and

for AX_2 and AX_3 type molecules having two and three valence-shell bond
pairs ($BeCl_2$ and BF_3, for example). Thus, the geometric arrangement of the
ligands about a central atom would be that which maximizes the interpair
separation, for bonded and non-bonded pairs. In this way, electrons of equal
spin will be as far apart as possible.

By additionally proposing that a domain for a non-bonding pair is larger
and broader, for example

for NH_3, the differences in the bond angles on going from CH_4 to NH_3 and
H_2O can be interpreted. The geometries of ClF_3, XeF_3^+ and XeF_2 can also
be reproduced:

In these molecules, the non-bonding electron pairs of the central atom are in equatorial positions where there is more space.

Two other assumptions which rationalize small changes in molecular geometry are: (a) a multiple bond is regarded as a single domain, although of larger size than a single-bond electron pair domain; and (b) bonding pair domains in the valence shell of a central atom decrease in size with increasing electronegativity of the ligand, and increase in size with increasing electronegativity of the central atom.

Problem 8.4 Confirm that the geometries below are in agreement with the VSEPR model:

(a) ICl_4^- *(square planar)*
(b) NO_2^- *(bent)*
(c) SF_6 *(octahedral)*
(d) NO_3^- *(trigonal planar).*

Problem 8.5 Predict the differences in bond angles in ethylene and in chloromethane with respect to $120°$ *and* $109.5°$, *respectively.*

The VSEPR model shows, therefore, a very high *predictive power* which, when coupled with its apparent simplicity, accounts for its great popularity. However, if high predictive capacity is a necessary condition for the acceptance of any theoretical model, it is not sufficient. Indeed, it could be argued that, strictly speaking, the VSEPR model lacks sufficient *explanatory power*, because it ignores other important contributions to the total molecular energy, for instance the Coulomb energy of electrons and nuclei and (at least in part) the electron exchange energy, as well as changes in the kinetic energies of the electrons. It could of course be said that the model just happens to pick the main contribution to the differences in energy for various geometric configurations of the nuclei. Thus, one is faced with the following dilemma. On one hand, it is difficult to accept that the changes in the Coulomb components of the energy with geometry are unimportant. If the electrostatic repulsion between the electron pairs was included, reinforcing the Pauli repulsion, the electron–nucleus attractions – the very ones that keep the molecules as a stable collection of atoms – would still be left out. On the other hand, it is equally difficult to admit that such a success in correctly predicting gross molecular geometries and interpreting small differences is purely accidental.

A more fundamental criticism of VSEPR as an explanatory model is that it relies upon a consideration of localized pairs of electrons having opposite spins. However, as already mentioned in our discussion of $\rho(\mathbf{r})$, localized

pairs of electrons, bonding or non-bonding, are not evident in the topological properties of the charge density. For example, for a tetrahedral AX_4 molecule the total electron cloud is more correctly represented as (a) than as (b):

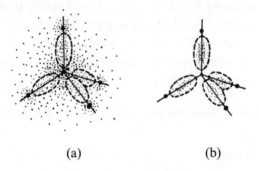

(a)   (b)

and thus electron pair repulsions become more difficult to characterize. Moreover, as shown by Bader (ref. 91), electron pair density does not, in general,
define regions of space beyond the atomic core in which pairs of electrons
are localized.

It is quantum mechanically acceptable to consider that, for two valence
electrons A and B, the likelihood of spatial arrangements decreases in the
following order

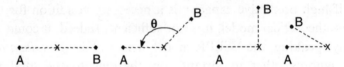

This means that, for a given instantaneous position of reference electron A,
electron B will be more likely to be found in the opposite hemisphere. This,
however, does not entitle one to consider that the two electrons occupy
opposite domains:

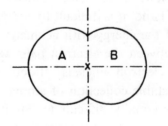

(see also ref. 108).

If the work of Bader was invoked above to criticize VSEPR as an
explanatory model, it is indeed the same author who offers some physical
interpretation of the high predictive power of the method. This is to be found

not in the topology of the charge density $\rho(\mathbf{r})$ but in the corresponding second derivatives, the laplacian of $\rho(\mathbf{r})$. We have already introduced the idea of regions where the function $\rho(\mathbf{r})$ is locally concentrated (or depleted), in a sense that does not imply the existence of maxima (or minima) in the charge density. Thus, in CH_4 there are four tetrahedrally arranged concentrations of charge, in the sense above. Similarly, the oxygen atom of the H_2O molecule exhibits four local concentrations of valence-shell charge: two bonded concentrations and two larger and equivalent non-bonded concentrations. The parallel with the VSEPR model is tempting in that the localized pairs of electrons in this model can be thought of as corresponding to the local concentrations of electronic charge around the central atom which are a result of the ligand field operating in concert with the Pauli principle. It has indeed been demonstrated that the total energy of a molecule is minimized for the geometry which maximizes the separations between the valence-shell charge concentrations defined in terms of the laplacian of $\rho(\mathbf{r})$ (ref. 109).

8.6 Canonical molecular orbitals and localized functions

As we have seen previously (Chapter 5), the eigenfunction for a polyelectronic atom is antisymmetric with respect to the exchange of the coordinates of any two electrons, and can be expressed as a Slater determinant whose elements are the various occupied spin-orbitals (or a linear combination of Slater determinants, in the case of open-shell atoms). The same applies to polyelectronic molecules, the atomic orbitals being replaced by the various occupied molecular orbitals associated with the α and β spin-functions: spin molecular orbitals. Thus, for the molecules H_2O, NH_3 or CH_4 having five doubly occupied m.o.s (one core s orbital and four valence m.o.s), we have

$$\Psi = \frac{1}{\sqrt{5!}} \begin{vmatrix} \psi_1(1) & \psi_1(2) & \cdots & \psi_1(10) \\ \bar{\psi}_1(1) & \bar{\psi}_1(2) & \cdots & \bar{\psi}_1(10) \\ \psi_2(1) & \psi_2(2) & \cdots \\ \bar{\psi}_2(1) & \cdots \\ \vdots \\ \bar{\psi}_5(1) & \cdots \end{vmatrix} \qquad (8.15)$$

with energies ε_i corresponding to the m.o.s ψ_i (all different ε_i values in the case of H_2O, and two and three degenerate m.o.s for NH_3 and CH_4, respectively).

The total electronic energy (see Eq. (7.26)) is given by

$$E = \sum_i^5 \left[2\varepsilon_i - \sum_j^5 (2J_{ij} - K_{ij}) \right].$$ (8.16)

The total electron density function is given by

$$\Psi^* \Psi = 2 \sum_{i=1}^5 \psi_i^* \psi_i$$ (8.17)

if the m.o.s are all orthogonal (if not, there would be cross-contributions $\psi_i^* \psi_j$).

The four valence m.o.s for H_2O are schematically shown in Fig. 8.7 (see also Chapter 7) and are related to the energy diagram of Fig. 7.7 and to the photoelectronic spectrum of Fig. 7.8. The various orbitals can be named a_1, b_1 and b_2 according to how they behave under each symmetry operation involving the twofold z rotation axis and the nuclear yz symmetry plane:

	Rotation around the z axis	Reflection in the yz plane
m.o.s a_1	Unchanged	Unchanged
m.o. b_1	Sign change	Sign change
m.o. b_2	Sign change	Unchanged

Orbital $1a_1$ is almost entirely associated with the oxygen 2s orbital, hence it is practically a non-bonding m.o. There is another non-bonding m.o. of π symmetry ($1b_1$) and two bonding m.o.s ($1b_2$ and $2a_1$).

The existence of two non-bonding electron pairs and two bonding electron pairs is in accordance with the classical structure

but there is no correspondence between each formal bond and each bonding m.o. Each bonding m.o. involves simultaneously the two H atoms in equal terms, that is, the two OH bonds are equivalent but not independent.

The lack of correspondence between formal bonds and bonding molecular orbitals is a general feature for polyatomic molecules, but this should not be considered as a problem, because, strictly speaking, each m.o. has no

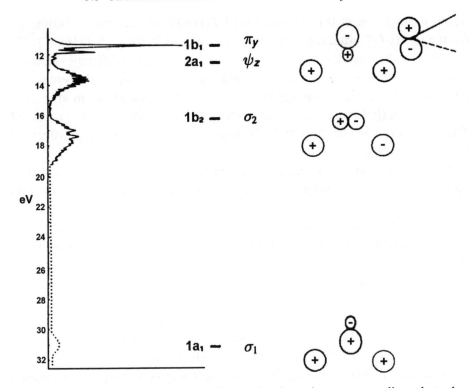

Fig. 8.7 Occupied valence m.o.s of H_2O related to the corresponding photoelectronic spectrum. (See also ref. 110.)

physical meaning by itself. Of course m.o.s and the corresponding energy values constitute a rich means of interpreting many molecular properties, but in quantitative terms m.o.s serve only as a basis for the definition of the eigenfunction Ψ, which is the proper way to describe the physical reality *viz.* the global electronic structure of a molecule. It is true that there is one property which can be directly associated, in an approximate manner, with the m.o.s ψ_i: it is the experimental ionization energies I_i which, according to Koopmans' theorem and as we have seen in Chapters 5 and 6, are approximately equal to $-\varepsilon_i$. This is illustrated in Fig. 8.7 for H_2O. It should be noted, however, that the ionization energies are the energy change for the ionization

$$H_2O \rightarrow H_2O^+(k) + e^-(k) \tag{8.18}$$

that is,

$$I_k = E(H_2O^+(k)) - E(H_2O) \tag{8.19}$$

which depends on Ψ for H_2O (through $E(H_2O)$) and on the eigenfunction for the ion H_2O^+ obtained by removal of electron k, through $E(H_2O^+)$. The various I_k values are the result of obtaining H_2O^+ ions in different electronic states depending on the k electron removed: H_2O^+ in its fundamental configuration, if the highest energy electron is removed, or H_2O^+ in an excited configuration (holes in m.o.s of lower energy), if another electron is removed. Therefore, even in the case of ionization potentials it is the eigenfunctions that really matter.

To consider that it is Ψ and not the form of each individual m.o. that ultimately matters is a consequence of the determinantal form of the wavefunction. In fact, determinants remain unchanged when the respective elements are subject to some specific operations (unitary transformations) as was already illustrated on page 89 (Problem 5.2). For example, the eigenfunction (Eq. (8.15)) is not changed if we replace each m.o. (row i) by $\chi_i = \psi_i + \lambda\psi_j$. Such mathematical alterations 'inside' the determinant do not correspond to any physical change, because the eigenfunction Ψ remains unchanged.

It is important to note, however, that the new functions χ_i cannot be directly related to the ionization energies (except if ψ_i and ψ_j are degenerate orbitals, see page 20). The functions ψ_i are eigenfunctions of a hamiltonian, in particular, the Hartree–Fock hamiltonian (Chapter 5); hence they are associated with energy values ε_i. But, in general, the functions χ_i *are not eigenfunctions of any hamiltonian*. Therefore, it is not possible to speak of an energy corresponding to each of these functions. Only the functions ψ_i are *canonical molecular orbitals*; the functions χ_i cannot, rigorously, be called orbitals as this word is necessarily associated with the notion of energy. In other words, the functions ψ_i diagonalize the matrix of the hamiltonian $\hat{H}$ (Chapter 2), that is

$$\langle\psi_i|\hat{H}|\psi_i\rangle = \varepsilon_i$$
$$\langle\psi_i|\hat{H}|\psi_j\rangle = 0 \tag{8.20}$$

but the functions χ_i do not:

$$\langle\chi_i|\hat{H}|\chi_j\rangle \neq 0. \tag{8.21}$$

On the basis of a set of functions χ_i the total electronic energy is no longer given by an expression equivalent to Eq. (8.16), but its value

$$E = \langle\Psi|\hat{H}|\Psi\rangle \tag{8.22}$$

is independent of the set of functions (ψ_i or χ_i) which are used. Similarly, the total electron density function $\Psi^*\Psi$ remains the same. The molecule is oblivious to the use of a set of canonical m.o.s or a new set of functions derived from them by a unitary transformation.

Among the infinite number of ways of transforming the set of canonical m.o.s without changing the eigenfunction Ψ, some would conform better to the classical structural formula of the molecule, namely those unitary transformations that lead to χ_i functions which are practically localized in each formal bond (or in individual atoms, as electron lone pairs). For example, by replacing the canonical σ_1 and σ_2 molecular orbitals of H_2O by

$$\sigma_1 + \sigma_2$$
$$\sigma_1 - \sigma_2 \tag{8.23}$$

equivalent χ_1 and χ_2 functions are obtained which are mainly localized in the formal O–H(1) and O–H(2) bonds, that is, functions whose main contributions come from the O atomic orbitals and the 1s H(1) and 1s H(2) orbitals, respectively (Fig. 8.8).

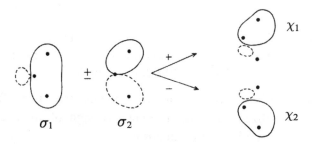

Fig. 8.8 Definition of χ_1 and χ_2 functions almost localized in the formal O–H(1) and O–H(2) bonds of H_2O.

A similar discussion will be presented for CH_4. The set of four canonical valence m.o.s of CH_4 (see Chapter 7 and refs. 111 and 112) all involve simultaneously the $1s_j$ orbitals of the four H atoms, alongside the s and the 2p (x, y and z) valence atomic orbitals of C:

$$\psi_1 = 0.58s + 0.19(s_1 + s_2 + s_3 + s_4)$$
$$\psi_2 = 0.55x + 0.32(s_1 - s_2 + s_3 - s_4)$$
$$\psi_3 = 0.55y + 0.32(s_1 + s_2 - s_3 - s_4) \tag{8.24}$$
$$\psi_4 = 0.55z + 0.32(s_1 - s_2 - s_3 + s_4).$$

From these m.o.s a set of four χ_i functions can be obtained, each one mainly localized in each C–H formal bond. The way to obtain χ_i from ψ_i is to use the same linear combination seen in Eq. (8.24) for the 1s H orbitals (including a normalization factor $1/2$):

$$
\begin{aligned}
\chi_1 &= (\psi_1 + \psi_2 + \psi_3 + \psi_4)/2 \\
&= 0.29s + 0.28(x + y + z) + 0.58s_1 - 0.07(s_2 + s_3 + s_4) \\
\chi_2 &= (\psi_1 - \psi_2 + \psi_3 - \psi_4)/2 \\
&= 0.29s + 0.28(-x + y - z) + 0.58s_2 - 0.07(s_1 + s_3 + s_4) \\
\chi_3 &= (\psi_1 + \psi_2 - \psi_3 - \psi_4)/2 \\
&= 0.29s + 0.28(x - y - z) + 0.58s_3 - 0.07(s_1 + s_2 + s_4) \\
\chi_4 &= (\psi_1 - \psi_2 - \psi_3 + \psi_4)/2 \\
&= 0.29s + 0.28(-x - y + z) + 0.58s_4 - 0.07(s_1 + s_2 + s_3).
\end{aligned}
\tag{8.25}
$$

Inspection of expressions (8.25) shows the following.

(a) For each function χ_i there is almost no contribution from the 1s orbitals of atoms H_j, with $j \neq i$.
(b) The functions χ_i are equivalent as far as the contributions of the 2s and 2p orbitals of C and the 1s orbital of H are concerned, but differ in their orientation as a result of the variable signs of the 2p contributions.

The small contribution of atoms H_j to χ_i means that these functions are not completely localized. Indeed, the negative sign of these contributions warrants the orthogonality of the functions.

Problem 8.6 *For each χ_i function in Eq. (8.25) calculate the ratio between the sum of the squared coefficients for the 2p orbitals and the square of the coefficient for the 2s orbital.*

It is found (Problem 8.6) that, for each χ_i, the global contribution of the 2p orbitals is approximately three times that of the 2s. That is, the 'mixing' of 2s and 2p C orbitals involved in the construction of equivalent and almost localized functions is approximately equivalent to the consideration of sp^3 hybrid orbitals. Thus, ignoring residual delocalizations, the functions χ_i can be represented as in Fig. 8.9.

This figure is identical to Fig. 8.6, which means that we have arrived at a result similar to the valence-bond description made previously. It was, in fact, only after the introduction of equivalent molecular orbitals (which later evolved into almost localized m.o.s) by J.E. Lennard-Jones in 1949 to provide the m.o. equivalent of directed valence bonding (ref. 113) that m.o. theory found widespread application to structural problems in chemistry.

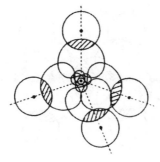

Fig. 8.9 Functions χ_i localized in each formal bond of CH_4.

Again, the functions corresponding to each formal bond are equivalent but not independent, that is

$$\langle \chi_i | \hat{H} | \chi_j \rangle \neq 0 \qquad (8.26)$$

because χ_i are not eigenfunctions of the hamiltonian $\hat{H}$. These functions are usually called localized orbitals. We keep this name but use inverted commas because no energy can be associated with them.

A similar study for H_2O (ref. 111), now including both bonding and non-bonding m.o.s, leads to two equivalent molecular 'orbitals' localized on O (which are approximately $sp^{2.5}$ hybrids) and to two equivalent molecular 'orbitals' localized at each O–H bond. The latter are, approximately, the result of the superposition of the 1s orbital of H with an sp^4 hybrid orbital of O (along with a small contribution from the 1s orbital of the other H atom). For an sp^4 orbital, the square of the coefficient of 2p is four times that for 2s.

Problem 8.7   Confirm that $0.45s + 0.89p$ represents an sp^4 hybrid orbital.

Problem 8.8   Interpret the structure of NH_3 in terms of approximate localized 'orbitals'.

The quasi-localized 'orbitals' of CH_4 imply sp^3 hybrid orbitals in so far as the molecular geometry is tetrahedral. The same occurs for ethane (Fig. 8.10) and all saturated unstrained hydrocarbons. Thus it is found that, as a result of an essentially tetrahedral arrangement of four atoms around each C, all C–H bonds in these compounds can be described in terms of identical localized molecular 'orbitals'. Although these cannot be directly related to electron energies, this description parallels the observation that C–H bond

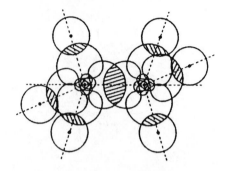

Fig. 8.10 Quasi-localized molecular 'orbitals' for C_2H_6.

energies and bond lengths do not change significantly from molecule to molecule. The same applies to C–C bonds.

Problem 8.9   Justify the fact that there are not just two ionization energies (for the valence shell) in the case of the ethane molecule, as the classical formula would seem to indicate:

$$
\begin{array}{c}
\text{H} \quad \text{H} \\
| \quad\quad | \\
\text{H}-\text{C}-\text{C}-\text{H} \\
| \quad\quad | \\
\text{H} \quad \text{H}
\end{array}
$$

For the molecule BeH_2, which has linear geometry, the appropriate hybrid orbitals of Be in the definition of the functions χ_i, quasi-localized in each formal Be–H bond, are two collinear sp hybrid orbitals (see Chapter 3):

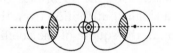

For the trigonal planar BH_3 molecule, the appropriate hybrid orbitals of B are three sp^2, at angles of 120°:

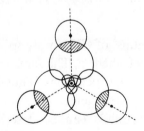

For further discussion see refs. 110, 114–116.

8.7 Use and misuse of the hybrid orbital concept

We have just seen that there is a *correspondence* between the geometry of molecules and the hybrid orbitals that are used to construct the appropriate quasi-localized molecular 'orbitals'. Assuming a known molecular geometry, one can identify these hybrid orbitals, if there is an advantage in replacing the initial set of canonical m.o.s by a set of localized molecular 'orbitals'. Since this mathematical substitution does not change the wavefunction Ψ, atomic orbital hybridization cannot be regarded as a physical phenomenon. Also, *no cause–effect relationship* can be established between hybridization and molecular geometry. On the contrary, it is from the knowledge of the geometry that, if convenient, hybrid orbitals can be mathematically defined (without any physical consequences).

It is in this light that many statements about molecular geometry, still encountered in teaching, must be considered to be incorrect or at least misleading. Some typical statements follow. 'The methane molecule has tetrahedral geometry, because the C atom is sp^3 hybridized'. 'sp hybridization leads to collinear bonds'. 'The complex $[Cr(NH_3)_6]^{3+}$ is octahedral due to d^2sp^3 hybridization of the chromium atom'. What can be said is that, if one wants to define quasi-localized molecular 'orbitals' for CH_4, then, because of the tetrahedral geometry, the orbital contribution of the C atom can be identified as being sp^3 for each bond. The use of loose expressions like 'an sp^3 carbon atom' to mean a C atom bonded to four other atoms in a tetrahedral arrangement has contributed to perpetuate the above misunderstanding.

It is stressed again that, assuming a given basis set of 1s orbitals for the H atoms and 2s and 2p orbitals for C, exactly the same wavefunction Ψ is obtained irrespective of using the canonical m.o.s ψ_i (as linear combinations of a.o.s) or the molecular 'orbitals' χ_i (linear combinations of ψ_i through a unitary transformation). Hence, for example the calculated molecular energy is exactly the same, provided that the residual interactions given by

$$\langle \chi_i | \hat{H} | \chi_j \rangle \neq 0 \tag{8.27}$$

are not ignored in the calculation. As a consequence, *no geometric parameter or any other molecular property can be explained by invoking hybrid orbitals*.

If all orbital interactions of the type described by Eq. (8.27) are included in the calculations, any result obtained is independent of the combination of a.o.s used, if any, i.e. four pure atomic orbitals, four sp^3 hybrid orbitals, three sp^2 and one p orbital, two sp and two p orbitals, etc. In particular, at a given level of approximation, the linear combinations of hybrid (and other) orbitals according to the variational principle studied in Chapter 7 lead to the same set of canonical m.o.s, as can be seen if each hybrid orbital in the expression for the m.o.s is replaced by the corresponding s and p components.

It is only when the interactions expressed by Eq. (8.27) are neglected that orbital hybridization *seems* to have an effect and, thus, *seems* to explain some molecular properties. This is so, both in m.o. theory and in v.b. theory. Ironically, it is more difficult to take these interactions into account in v.b. theory where the concept of hybrid orbital was first introduced. For example, it is more difficult to do a complete v.b. treatment of CH_4, with inclusion of the less traditional formulae in the resonance scheme (and which accounts for the non-independence of bonds), than to do a fully delocalized treatment by m.o. theory, at the same degree of approximation. But if the complete treatment is carried out, the v.b. result is the same irrespective of the hybridization considered, just as any m.o. result is the same whether pure, hybrid orbitals or a mixture of both are taken as the basis for the linear combinations.

Hybrid orbitals can, of course, be chosen so that the interactions of type (8.27) are minimized. That is why it is more appropriate, for example, to choose sp^3 hybrids for CH_4 and sp hybrids for C_2H_2 and not the other way round. In so doing, the calculated total energy and electron density distribution do not vary significantly whether or not such residual interactions are included in the calculations. Thus, those molecular properties that are directly related to the total energy or to the total electron density distribution can be directly interpreted in terms of the most localized m.o.s or v.b. functions. This is the case for the almost constancy of values of C–H, C–C, O–H, etc. bond energies in different compounds and for the interpretation of dipole moments in terms of independent bonds, as we have previously mentioned. But in other cases, even the small residual interactions are of paramount importance, in particular for those properties that depend directly on the energy levels of the electrons. For example, due to such residual interactions, the prediction of only one ionization energy for the valence shell of CH_4 on the basis of localized bonds fails. This subject will be taken up again and extended to include other properties in Chapter 12. Meanwhile, in the coming chapters, the description of the structure of some molecules involving π systems will be simplified by the use of localized σ molecular 'orbitals'.

At this stage, it is appropriate to draw attention to additional misuses of the concept of orbital hybridization, besides the relation with molecular geometry already discussed. We will briefly refer here to frequent interpretations of differences in bond energies, bond lengths, force constants, bond polarity, acidic character, etc. (see ref. 117).

It is not uncommon to attribute sp³ hybridization of the carbon atom in CH_4 to a resulting increase of the overlap integral with the H 1s orbitals, the claim being that this leads to stronger C–H bonds. The fact that the molecular energy does not depend upon the type of hybridization considered (if any) shows that such an interpretation is not valid. Molecular energy and bond energies depend on the geometry of the molecule and not on the orbital hybridization considered (if any). What can be said is that, for four tetrahedrally arranged CH bonds, the hybrid orbitals that lead to least residual delocalization interactions are the four equivalent sp³ hybrids.

It is also common to attribute the fact that the CH bond is shorter in C_2H_2 than in C_2H_6 to sp hybridization in the former case, as more s-character in the hybrid orbital implies a larger overlap integral. However, it should be noted that the smaller CH bond length and the consideration of sp carbon hybrids in acetylene in order to define quasi-localized molecular 'orbitals' are two manifestations (one physical, the other mathematical) of the same real feature: the number of atoms bonded to each C atom and their geometric arrangement.

Similarly, more s-character in overlapping hybrid orbitals usually agrees with the observation of stretching vibrations at higher frequency, but it cannot be taken as an explanation of such an observation. In addition, higher s-character of the C hybrid orbital for an almost localized C–H bond, in general, goes together with a larger polarity of the bond (greater effective electronegativity of C), larger acidic character, smaller magnetic shielding of the H nucleus, larger ^{13}C–H nuclear spin coupling constant, etc. However, it cannot be used as an explanation of these observations. *No cause–effect relationships* can be drawn in this way, but useful *correlations* can be established: correlations between the observations themselves and with the hybrid orbitals more appropriate to a localized bond description of C–H.

Another field where care must be exercised when using the hybrid orbital concept deals with 'bent' bonds and molecular strain. Cyclopropane is the best example. The molecular strain is usually justified by the occurrence of 'bent' CC bonds, and these are interpreted in terms of hybrid orbitals that cannot have axes making angles of 60° to coincide with the C–C nuclear axes (Fig. 8.11).

Firstly, it is stressed again that the molecular energy, even though it depends on the molecular geometry, does not depend on the hybrid orbitals considered (if any). In addition, it should be noted that the electron density in each CC section corresponds not only to the respective CC localized molecular 'orbital' but has also significant contributions from other orbitals, due to the geometry of the molecule. As a result, the deviations of electron

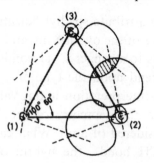

Fig. 8.11 Localized molecular 'orbital' for one of the CC bonds of cyclopropane.

density with respect to cylindrical symmetry around the CC nuclear axis cannot be entirely interpreted in terms of hybrid orbitals, as is sometimes implied. There is thus a difference between bent localized molecular 'orbitals' and bent bonds. It is with the latter that the bond stress of the molecule can be (in part) related. Bader has been able to reproduce the experimentally determined strain energy of cyclopropane from considerations of the total distribution of electron charge (ref. 118).

It has already been mentioned that hybrid orbitals can be chosen so as to minimize the residual interactions between localized molecular 'orbitals'. It is not uncommon for the hybrids chosen in this way to have axes that do not conform to the molecular geometry. For example, when it is said that the Cl atoms of chloroform increase the s-character of the C hybrid involved in the CH bond, an angle between the axes of the other three hybrid orbitals smaller than 109.5° is implied. In fact, the actual Cl–C–Cl bond angles are larger. Again, bent quasi-localized molecular 'orbitals' occur independently of any deviations of the electron density in each CCl sector with respect to cylindrical symmetry around the C–Cl nuclear axis.

In conclusion, models based on hybrid orbitals have been frequently extended beyond their valid use. Further examples are found in Chapters 11 and 12.

9

π Molecular orbitals: conjugation and resonance

9.1 The σ–π separation

In Chapters 4 and 6 we have distinguished between σ molecular orbitals and π molecular orbitals in diatomic molecules on the basis of the symmetry of the m.o.s with respect to rotation around the internuclear axis: whereas a σ orbital has cylindrical symmetry, a π orbital changes sign upon a rotation of 180°. In Chapters 7 and 8 we have extended the notion of σ orbitals to polyatomic molecules by referring to the local symmetry with respect to each X–Y internuclear axis. We will now study systems of π m.o.s in polyatomic species.

As before, the electronic wavefunction Ψ is constructed as a Slater determinant whose elements are the occupied spin-orbitals, both of σ and π symmetry. In the case of open-shell species or if configuration interaction is considered, then more than one determinant must be used.

The total electronic energy E (see Eq. (7.26)) is given by

$$E = \sum_i^{\text{occ.}} \left[2\varepsilon_i - \sum_j^{\text{occ.}} (2J_{ij} - K_{ij}) \right] \tag{9.1}$$

in terms of both sets, σ and π, of occupied m.o.s. Whenever for some purposes the total electronic energy of a molecule is taken as a simple sum of a σ component and a π component

$$E = E_\sigma + E_\pi \tag{9.2}$$

an approximation is being made, because repulsion and exchange involving any σ electron and any π electron cannot be ignored. The expression (9.2) would require that the wavefunction Ψ was expressed as a product of σ and π components

$$\Psi = \Psi_\sigma \times \Psi_\pi \tag{9.3}$$

the same being required of the hamiltonian

$$\hat{H} = \hat{H}_\sigma \times \hat{H}_\pi \tag{9.4}$$

and this is not possible without approximations. The effects of the approximation are expected to be smaller the larger the difference between the energies that can be associated with π and σ electrons separately.

Nevertheless, a separate discussion of σ and π molecular orbitals is possible and useful in many qualitative discussions and even whenever more rigorous quantitative energy considerations are needed provided that the interplay of σ and π electrons is taken into account, as is in fact necessary within either set of σ and π electrons.

9.2 The CO_2 molecule and the CO_3^{2-} ion

The study of these species will follow closely the cases of BeH_2 and BH_3 in Chapter 7, by taking advantage of the identical geometries: linear for CO_2 and BeH_2 and trigonal planar for CO_3^{2-} and BH_3.

The essential difference between the nature of the m.o.s in BeH_2 and in CO_2 is that the peripheral atoms in carbon dioxide (O atoms) contribute with four valence atomic orbitals and not just one as in BeH_2. However, the 2s a.o. of the oxygen atom is of sufficiently low energy relative to the 2p a.o.s (as well as relative to the carbon valence a.o.s) in order to be considered – to an acceptable approximation for many purposes – as a core orbital (besides the 1s orbitals of C and O). We thus have three (from O) + four (from C) + three (from O) = 10 a.o.s to define 10 valence m.o.s. Orthogonality relations associated with the symmetry of the various orbitals enable those 10 a.o.s to be divided into three sets as shown in Fig. 9.1.

The three 2p orbitals in both the first and second sets in Fig. 9.1, having parallel axes, define three π m.o.s. There will be three π_z and three π_y molecular orbitals as linear combinations of the $2p_z$ and the $2p_y$ orbitals of C and O. In order to arrive at the form of these functions, it is convenient to start by making the following intermediate combinations of the $2p_\pi$ oxygen orbitals (in-phase or symmetrical ϕ^s and out-of-phase or antisymmetrical ϕ^a):

$$\begin{aligned}
\phi_z^s &= z_{O(1)} + z_{O(2)} & \phi_y^s &= y_{O(1)} + y_{O(2)} \\
\phi_z^a &= z_{O(1)} - z_{O(2)} & \phi_y^a &= y_{O(1)} - y_{O(2)}.
\end{aligned} \tag{9.5}$$

Only the symmetrical combinations are not orthogonal with the carbon $2p_\pi$ a.o.s. We thus have two bonding π m.o.s given by (ignoring the normalization

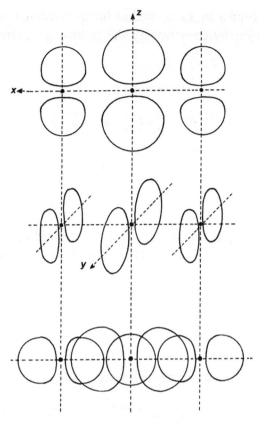

Fig. 9.1 The three sets of valence atomic orbitals for the linear arrangement O $\cdots$ C $\cdots$ O, according to an orthogonality criterion (the 2s a.o. of oxygen is taken as a core orbital).

factor for simplicity):

$$\pi_z = (z_{O(1)} + z_{O(2)}) + z_C$$
$$\pi_y = (y_{O(1)} + y_{O(2)}) + y_C, \tag{9.6}$$

two corresponding anti-bonding m.o.s

$$\pi'_z = (z_{O(1)} + z_{O(2)}) - z_C$$
$$\pi'_y = (y_{O(1)} + y_{O(2)}) - y_C, \tag{9.7}$$

and two non-bonding m.o.s involving only the O atoms

$$\pi_{zn} = z_{O(1)} - z_{O(2)}$$
$$\pi_{yn} = y_{O(1)} - y_{O(2)}. \tag{9.8}$$

Regarding the four σ m.o.s defined as linear combinations of $\chi_{O(1)}$, $\chi_{O(2)}$, χ_C and s_C, we begin by constructing the following intermediate combinations

$$\phi_1 = \chi_{O(1)} + \chi_{O(2)} \qquad (9.9)$$

and observing that ϕ_1 is orthogonal to the 2s orbital of C, but not to $2p_x$, whereas the opposite occurs with ϕ_2. Hence, for the two bonding m.o.s we have

$$\sigma_1 = as_C + b(\chi_{O(1)} - \chi_{O(2)})$$

$$\sigma_2 = c\chi_C + d(\chi_{O(1)} + \chi_{O(2)}) \qquad (9.11)$$

and for the anti-bonding m.o.s

$$\sigma_1' = a's_C - b'(\chi_{O(1)} - \chi_{O(2)})$$
$$\sigma_2' = c'\chi_C - d'(\chi_{O(1)} + \chi_{O(2)}). \qquad (9.12)$$

The m.o. diagram for CO_2 is thus as shown in Fig. 9.2, considering that the σ bonding orbitals have lower energy than the π ones and that σ_1 is the lowest in energy, as the energy associated with the 2s carbon orbital is lower than 2p.

The valence electronic configuration for this molecule (16 valence electrons) is thus

$$(s_{O(1)})^2, (s_{O(2)})^2, (\sigma_1)^2, (\sigma_2)^2, (\pi_y)^2, (\pi_z)^2, (\pi_{yn})^2, (\pi_{zn})^2 \qquad (9.13)$$

which means four non-localized bonding electron pairs and four non-bonding electron pairs associated with the O atoms. The average bond order is 2 and the

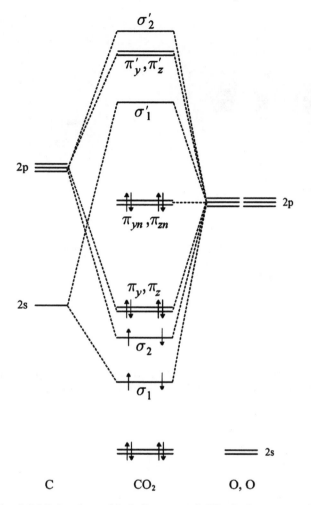

Fig. 9.2 Molecular orbital diagram of CO_2 (valence m.o.s).

classical formula is

$$\overline{O}=C=\overline{O}$$

The contour plots for the various m.o.s are shown in Fig. 9.3 and the contour map of the total electronic charge distribution is displayed in Fig. 9.4.

The canonical bonding m.o.s σ_1 and σ_2 can be combined adequately (unitary transformation) to yield two quasi-localized molecular 'orbitals' in

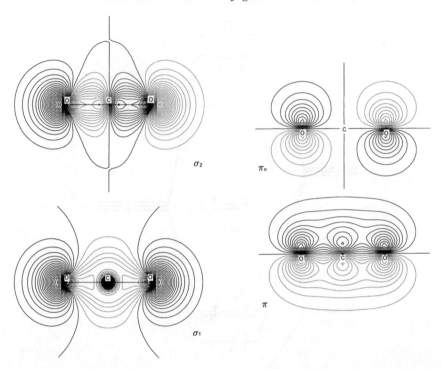

Fig. 9.3 Isocontours for the valence m.o.s of CO_2 (the two pairs of degenerate π m.o.s are represented by one orbital each, and the 2s contribution of the O atoms is ignored).

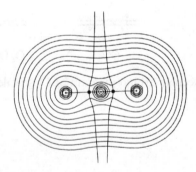

Fig. 9.4 Contour map of the ground state electronic charge distribution of CO_2 showing the positions of the interatomic surfaces defined by Bader (ref. 71) (see Chapter 8).

the sense encountered in Chapter 8: one almost entirely localized in O(1)C (negligible contribution of $x_{O(2)}$) and an equivalent one almost entirely localized in CO(2) (negligible contribution of $x_{O(1)}$). Reasonable approximations of these almost localized functions are defined in terms of two sp hybrid

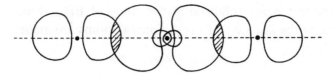

Fig. 9.5 Representation of the localized σ molecular 'orbitals' of CO_2 (the 2s orbital of O is taken as a core a.o.).

orbitals of the carbon atom:

$$\chi_1 = sp_1 + \chi_{O(1)}$$
$$\chi_2 = sp_2 + \chi_{O(2)}$$

(9.14)

and represented in Fig. 9.5.

We note, as in Chapter 8, that these localized functions are not canonical molecular orbitals (no energy can be associated with them) because they are not independent of each other, that is the interaction integral

$$\langle \chi_1 | \hat{H} | \chi_2 \rangle$$

(9.15)

does not vanish.

Problem 9.1

(a) *Using Eq. (9.14) expand the integral to find that one of the terms of integral (9.15) is* $\langle sp_1 | \hat{H} | sp_2 \rangle$.

(b) *Taking* $sp_1 = s + p$ *and* $sp_2 = s - p$ *(ignoring normalization factors), confirm that the term above is* $\langle s | \hat{H} | s \rangle - \langle p | \hat{H} | p \rangle$ *and that this difference cannot be zero.*

We turn now to the carbonate ion. Again, symmetry considerations as a consequence of the trigonal planar geometry of CO_3^{2-} enable the definition of a set of π molecular orbitals (four altogether) as linear combinations of the four 2p atomic orbitals (one from each atom) which have axes perpendicular to the nuclear plane (xy):

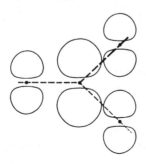

Similar to the procedure followed in the study of BH_3, three intermediate combinations of the $2p_z$ a.o.s of the peripheral atoms are made:

$$\phi_1 = z_{O(1)} + z_{O(2)} + z_{O(3)}$$
$$\phi_2 = z_{O(1)} - z_{O(3)} \qquad\qquad\qquad (9.16)$$
$$\phi_3 = z_{O(1)} - 2z_{O(2)} + z_{O(3)}.$$

It can be easily seen that ϕ_2 and ϕ_3 are orthogonal to z_C.

Problem 9.2   Confirm that ϕ_3 and z_C are orthogonal.

Thus, ϕ_2 and ϕ_3 are delocalized π non-bonding m.o.s and there is one delocalized π bonding m.o. given by

$$\pi_1 = az_C + b(z_{O(1)} + z_{O(2)} + z_{O(3)}) \qquad\qquad (9.17)$$

and the corresponding anti-bonding m.o.

$$\pi_1' = a'z_C - b'(z_{O(1)} + z_{O(2)} + z_{O(3)}). \qquad\qquad (9.18)$$

If we assume that the 2s orbitals of the oxygen atoms belong to the atomic core, then there will still be 3×2 (from the O atoms) $+ 3$ (from the C atom) $= 9$ a.o.s to use in the definition of nine m.o.s. These m.o.s have a mixed σ, π character, because they involve p orbitals with various orientations relative to the internuclear axes. That is the case, for example, of the m.o. constructed from the following 2p orbitals:

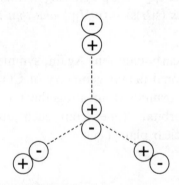

Note that the orientation of the oxygen 2p orbitals was chosen to coincide with the O–C axes. A more detailed analysis would show that three of the nine m.o.s are approximately non-bonding, there being three bonding and three anti-bonding ones.

The 24 valence electrons of CO_3^{2-} fill the four bonding m.o.s and the eight non-bonding ones (the 2s a.o.s of the O atoms being included). The average bond order is 4/3: each of the four equivalent CO bonds is between a single and a double bond.

Among the 12 occupied m.o.s, the three π m.o.s – one bonding and two non-bonding – are not only *non-localized* but also *non-localizable*. That is, it is not possible to make a unitary transformation with them that leads to three localized molecular 'orbitals'.

Problem 9.3 *Try to combine the orbitals* π_1, ϕ_2 *and* ϕ_3 *to conclude that it is not possible to obtain three* χ *functions localized in just one atom or two neighbouring atoms.*

On the contrary, out of the nine remaining m.o.s, three are localized in the O atoms (2s orbitals) and the other six (three bonding and three non-bonding) can be transformed into six quasi-localized 'orbitals'. In an approximate description, the three quasi-localized bonding 'orbitals', having σ local symmetry, can be defined in terms of the overlap of sp^2 hybrid orbitals of C with a 2p orbital of each O atom:

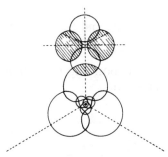

The three non-bonding ones are approximately 2p a.o.s of O whose axes are perpendicular to the internuclear CO axes.

These considerations and the non-localizability of the π orbitals are in accordance with the classical structure of CO_3^{2-} as a resonant hybrid:

In particular, the two π bonding electrons are equally distributed among the three OC sectors. Figure 9.6 shows a computer-generated relief map of the electron density of CO_3^{2-} taken from ref. 119.

Problem 9.4

(a) *Interpret the molecular structures of* O_3 *and* NO_2^- *in terms of localized 'orbitals' and non-localizable* π *orbitals.*

(b) *Confirm that the respective geometries are correctly predicted by the valence-shell electron pair repulsion model.*

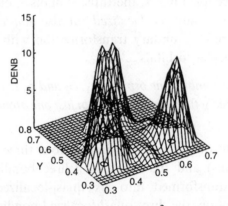

Fig. 9.6 Relief map of the electron density of CO_3^{2-} in the nuclear plane (reprinted with permission from ref. 119, *Journal of Chemical Education*, **70**, A76 (1993); copyright 1993, Division of Chemical Education Inc.).

9.3 The ethylene and acetylene molecules

The valence electronic structure of the ethylene molecule, C_2H_4, can be described in terms of a set of σ bonding molecular 'orbitals' approximately localized in the CC and CH bonds

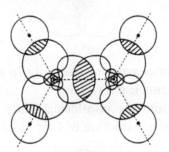

each one filled with two electrons, and two π canonical molecular orbitals: a bonding orbital filled with two electrons

$$\pi_z = z_1 + z_2 \tag{9.19}$$

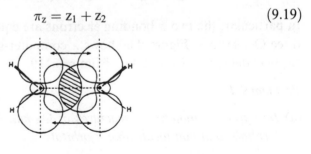

and the corresponding anti-bonding one, vacant,

$$\pi'_z = z_1 - z_2 \tag{9.20}$$

(normalization factors are being ignored in Eqs. (9.19) and (9.20)).

The geometry of the molecule points to the involvement of approximately sp^2 hybrid orbitals of the C atoms in the description of the localized 'orbitals'. The double bond

helps to make the planar geometry specially stable relative to a twisted geometry:

In particular, for a dihedral angle of 90° the 2p carbon orbitals responsible for the π bond would become orthogonal to each other. Accordingly, the energy barrier for internal rotation around the CC nuclear axis is especially high. This explains the occurrence of *cis–trans* isomerism in ethylene derivatives, a situation which does not happen in ethane where, for any torsion angle, the CC bond remains a single bond, thus enabling almost free internal rotation about the CC axis.

The higher energy of the π m.o. relative to the σ framework means that the bond energy for a C=C is less than twice the C–C bond energy. This contributes to the exothermicity, hence to the likelihood of addition reactions such as,

$$C_2H_4 + H_2 \rightarrow C_2H_6$$

where the π bonding m.o. together with the bonding m.o. of H_2 are replaced by two σ 'orbitals' associated with two additional C–H bonds.

One of the unitary transformations of the m.o.s which leaves the electronic wavefunction unaltered is the combination of the π m.o. and the σ 'orbital' localized in the CC bond, χ_{CC}:

$$\pi + \chi_{CC} \qquad\qquad \pi - \chi_{CC}. \tag{9.21}$$

This leads to two equivalent molecular 'orbitals' of mixed σ, π character, which are sometimes called 'banana bonds':

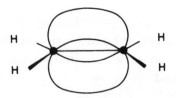

As mentioned in Chapter 8 (page 172), the double bond is associated with an elliptical distribution of electronic charge in the plane perpendicular to the CC nuclear axis and containing its mid-point where the electronic density has a local maximum (critical point in the theory of Bader). The relief diagram and the contour plots of Fig. 9.7 taken from the work of Bader *et al.* (ref. 92) show the distribution of the electronic charge density in the nuclear plane of the molecule.

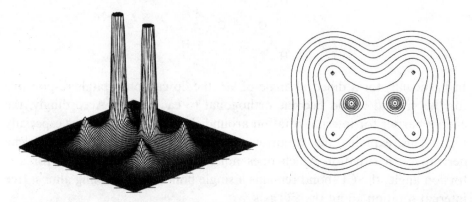

Fig. 9.7 Relief map and contour plots of the electron density of C_2H_4 in the nuclear plane (adapted with permission from ref. 92; copyright 1981, Institute of Physics Publishing).

Problem 9.5   Relate the structures of the isoelectronic molecules methanal and ethylene including a comparison of the corresponding molecular orbital description.

Problem 9.6   Extend to the π m.o.s of ethylene the calculation made on page 78 for the energies of the bonding and anti-bonding m.o.s of H_2^+ and conclude that the $\pi \to \pi'$ excitation energy is, with this approximation and ignoring the normalization factor, given by 2β, with $\beta = \langle z_1 | \hat{H} | z_2 \rangle$.

For the acetylene molecule, C_2H_2, we have, similarly, a framework of σ 'orbitals' approximately localized in the CC and CH bonds – which can be defined in terms of the direct overlaps involving the 1s orbitals of H and two sp hybrid orbitals of each C atom – plus four π m.o.s, two bonding, filled, and two anti-bonding, vacant:

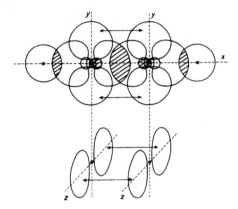

This description is in agreement with the classical formula

$$H-C\equiv C-H.$$

Problem 9.7   Relate the structures of the isoelectronic molecules acetylene and nitrogen including a comparison of the corresponding molecular orbital description.

9.4 The butadiene molecule

The molecule 1,3-butadiene, C_4H_6,

$$CH_2=CH-CH=CH_2$$

exists mainly as the *trans* isomer of planar geometry

the central CC bond being longer than the terminal CC bonds: 148 pm and 134 pm, respectively. The latter value is close to the one in ethylene (133 pm) but the former is significantly less than the typical value for a single C–C bond (154 pm in ethane). This suggests a bond order greater than 1 for the central CC bond.

The σ framework is a clear extension of what was considered for ethylene. Let us then consider the four π m.o.s as linear combinations of the four 2p carbon orbitals having axes perpendicular to the plane of the nuclei (the xy plane):

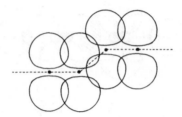

It is clear that there is overlap not only in each pair z_1, z_2 and z_3, z_4 but also between z_2 and z_3. Recalling the study of a linear H_3 molecule (page 141), we can relate the successive energies of the four π m.o.s to the number of nodes (besides the angular nodes of the p orbitals) in various combinations. Thus, the π orbital of minimum energy must be bonding for each pair of adjacent atoms (no new nodes):

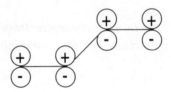

The next linear combination (next in energy) must have one node. The symmetry of the system then requires that this m.o. will be bonding in C(1)–C(2) and in C(3)–C(4) and anti-bonding in C(2)–C(3):

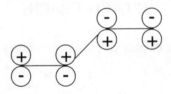

The next one will be anti-bonding in C(1)–C(2) and in C(3)–C(4) and bonding in C(2)–C(3) (two nodes):

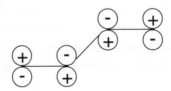

The π m.o. of highest energy will be anti-bonding in all CC sectors (three nodes):

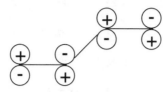

As only the two of lower energy are occupied (four electrons), the terminal CC bonds are necessarily stronger than the central CC bond. In fact, should the bonding and anti-bonding effects in C(2)–C(3) cancel each other out, a single C–C bond would be expected. But this can only be found through a proper calculation.

In order to ascertain the relative involvement of the 2p orbitals in the four π m.o.s,

$$\pi = \sum_{n=1}^{4} c_n z_n \tag{9.22}$$

it is necessary to use the variation principle already considered on page 157. This assumes the approximation of σ, π separation, namely that it is possible to minimize the energy of the π sub-system independently of the σ one. So, we will have to solve the secular determinant (page 161)

$$|H_{mn} - ES_{mn}| = 0 \tag{9.23}$$

in order to obtain the energies of the various m.o.s and, next, the coefficients c_n.

The secular determinant acquires a simple form if the following approximations are made,

(a) All H_{mm} integrals are taken as equal: common value α (Hückel Coulomb integral).
(b) All H_{mn} integrals are taken as equal for adjacent orbitals – common value β (Hückel resonance integral) – and zero otherwise.
(c) All overlap integrals S_{mn} are taken as zero and $S_{mm} = 1$ (normalized 2p orbitals).

The Coulomb integral α can be approximately equated to the energy of a carbon 2p electron. The neglect of overlap integrals is in part justified by a smaller value for two 2p orbitals of parallel axes than for collinear axes; however a typical value is still around 0.25. Resonance integrals all equal for adjacent orbitals means that the effect of differences in bond lengths are being neglected.

We then have:

$$
\begin{array}{c}
 & \begin{array}{cccc} z_1 & z_2 & z_3 & z_4 \end{array} \\
\begin{array}{c} \pi_1 \\ \pi_2 \\ \pi_3 \\ \pi_4 \end{array} &
\begin{vmatrix}
\alpha - E & \beta & 0 & 0 \\
\beta & \alpha - E & \beta & 0 \\
0 & \beta & \alpha - E & \beta \\
0 & 0 & \beta & \alpha - E
\end{vmatrix} = 0.
\end{array}
\tag{9.24}
$$

By dividing all elements of the determinant by β and making the substitution

$$(\alpha - E)/\beta = x \tag{9.25}$$

we get

$$
\begin{vmatrix}
x & 1 & 0 & 0 \\
1 & x & 1 & 0 \\
0 & 1 & x & 1 \\
0 & 0 & 1 & x
\end{vmatrix} = 0
\tag{9.26}
$$

which, when expanded, becomes

$$x^4 - 3x^2 + 1 = 0. \tag{9.27}$$

The roots of this equation are $x = \pm 0.62$ and $x = \pm 1.62$. By substituting into Eq. (9.25), the following energy values are obtained in increasing order of magnitude (note that β is a negative quantity)

$$
\begin{aligned}
E_1 &= \alpha + 1.62\beta \\
E_2 &= \alpha + 0.62\beta \\
E_3 &= \alpha - 0.62\beta \\
E_4 &= \alpha - 1.62\beta
\end{aligned}
\tag{9.28}
$$

for the four π orbitals (two bonding and two anti-bonding).

In order to calculate the coefficients c_n for each m.o. we need to construct the secular equations (see page 160):

$$
\begin{aligned}
\text{for } m = 1 \quad & xc_1 + c_2 = 0 \\
\text{for } m = 2 \quad & c_1 + xc_2 + c_3 = 0 \\
\text{for } m = 3 \quad & c_2 + xc_3 + c_4 = 0 \\
\text{for } m = 4 \quad & c_3 + xc_4 = 0.
\end{aligned}
\tag{9.29}
$$

Let us consider the lowest energy m.o. π_1. Then $x = -1.62$ and the Eq. (9.29) yield

$$c_2 = 1.62c_1$$
$$c_3 = -c_1 + 1.62c_2 = 1.62c_1$$
$$c_3 = 1.62c_4 \tag{9.30}$$
$$c_2 = c_3$$
$$c_1 = c_4.$$

By introducing the normalization condition, which for $S_{mn} = 0$ is

$$c_1^2 + c_2^2 + c_3^2 + c_4^2 = 1 \tag{9.31}$$

the individual c_n values are obtained and, finally,

$$\pi_1 = 0.37z_1 + 0.60z_2 + 0.60z_3 + 0.37z_4. \tag{9.32}$$

In an analogous way, the expressions for the other m.o.s are obtained:

$$\pi_2 = 0.60z_1 + 0.37z_2 - 0.37z_3 - 0.60z_4$$
$$\pi_3 = 0.60z_1 - 0.37z_2 - 0.37z_3 + 0.60z_4 \tag{9.33}$$
$$\pi_4 = 0.37z_1 - 0.60z_2 + 0.60z_3 - 0.37z_4.$$

It can already be concluded from the values of the coefficients that, in the region C(2)–C(3), π_1 is more bonding than π_2 is anti-bonding. This result helps us to understand the observation that the C(2)–C(3) bond length is shorter than the typical value for a single-bond.

Problem 9.8   Arrive at the normalization condition (9.31).

Problem 9.9   Obtain the expression for π_2.

Problem 9.10   Confirm that π_1 is normalized and orthogonal to π_2.

It is noted that the π m.o.s of butadiene, like those of CO_3^{2-} studied before, are not only non-localized but also non-localizable, that is there is no mathematical way of transforming them into an equivalent set of localized molecular 'orbitals'.

We can now compare the sum of the energies of the four π electrons of butadiene described by the non-localized π_1 and π_2 m.o.s

$$2 \times (\alpha + 1.62\beta) + 2 \times (\alpha + 0.62\beta) = 4\alpha + 4.48\beta \tag{9.34}$$

with the value that would be obtained if we had considered two localized π bonds as suggested by the classical structural formula. In this latter case, the sum would be four times the energy of the π bonding orbital of ethylene (see Problem 9.6), that is

$$2 \times 2 \times (\alpha + \beta) = 4\alpha + 4\beta. \tag{9.35}$$

The difference 0.48β is negative which means an additional stabilization of the butadiene molecule associated with the non-localization of the π molecular orbitals. This is a consequence of an interaction between the $2p_z$ a.o.s of atoms $C(2)$ and $C(3)$ which prevents the mere existence of two independent π bonds localized in $C(1)C(2)$ and $C(3)C(4)$. In this and similar situations one refers to *conjugation* of double bonds, a phenomenon clearly related to the molecular topology, that is to the sequence and geometric arrangement of atoms in the molecule.

We have referred above to the sum of the energies of the π electrons and not to the total π electron energy (within the approximation of σ, π separation). The difference is that in the former all repulsion and exchange energies are counted twice (see page 158). The total π electron energy would equal the sum of the energies of the π electrons only if the electrons could be treated as totally independent and distinguishable particles. Therefore, the stabilization energy associated with conjugation can only be equated to 0.48β in the simple model above if repulsion and exchange energies are not sensitive to delocalization.

The method used above for calculating the π m.o.s was originally proposed by the German physicist Erich Hückel (1896–1980), in 1930 (ref. 120). In spite of the approximations made, the *Hückel method* has had an appreciable success for two main reasons. One lies at its essentially topological basis. The other is that part of the approximations are accommodated by taking the α and β integrals as parameters adjusted by comparison with experiment. In this way the Hückel method becomes a *semi-empirical method*. However, modern computers enable some of the approximations to be easily removed, for example including all overlap integrals in the calculations. It is found that, by including overlap, the energies of the bonding m.o.s become closer, the opposite happening to the anti-bonding m.o.s, but the overall sequence is not usually altered.

Although the method was originally devised to deal with the conjugated π systems of organic molecules, in its numerical form it has also been applied to σ molecular orbitals, eventually including long-range resonance integrals H_{mn}: *extended Hückel calculations*. However, if the advantage provided by the simple analytical features of the Hückel method is to be replaced by more accurate numerical calculations, then the widely available program packages based on more reliable semiempirical methods (see page 162) are clearly preferable to the Hückel-type calculations.

9.5 The benzene molecule

In the case of the benzene molecule, C_6H_6, we consider 24 σ molecular orbitals of which 12 are bonding. These can be transformed into a set of quasi-localized molecular 'orbitals' which, according to the geometry of the molecule, can be defined approximately in terms of sp^2 hybrid orbitals of the C atoms together with the 1s orbitals of the H atoms:

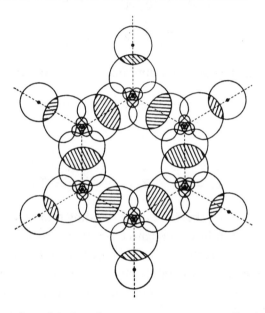

The six 2p atomic orbitals whose axes are perpendicular to the nuclear plane (the xy plane)

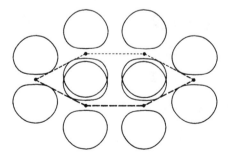

are linearly combined to define six π m.o.s. Although the high symmetry of the molecule permits the forms of the π orbitals to be established without solving the secular determinant (compare with the case of butadiene and note that, for example, z_1 interacts both with z_2 and with z_6)

$$\begin{vmatrix} x & 1 & 0 & 0 & 0 & 1 \\ 1 & x & 1 & 0 & 0 & 0 \\ 0 & 1 & x & 1 & 0 & 0 \\ 0 & 0 & 1 & x & 1 & 0 \\ 0 & 0 & 0 & 1 & x & 1 \\ 1 & 0 & 0 & 0 & 1 & x \end{vmatrix} = 0 \qquad (9.36)$$

it can be found that the corresponding polynomial equation

$$x^6 + 6x^4 + 9x^2 - 4 = 0 \qquad (9.37)$$

has the roots

$$\begin{aligned} x_1 &= -2 & x_2 &= -1 & x_3 &= -1 \\ x_4 &= 1 & x_5 &= 1 & x_6 &= 2. \end{aligned} \qquad (9.38)$$

It is immediately seen that there are three bonding m.o.s (three negative values of x_i) and three anti-bonding ones (three positive x_i roots). Furthermore, two bonding m.o.s are degenerate, the same being true of two anti-bonding m.o.s. The occurrence of degenerate orbitals raises a new problem, because any linear combination of them is equally acceptable (see page 20). A convenient choice leads to the following set:

$$\begin{aligned} \pi_1 &= (z_1 + z_2 + z_3 + z_4 + z_5 + z_6)/\sqrt{6} & E_1 &= \alpha + 2\beta \\ \pi_2 &= (z_2 + z_3 - z_5 - z_6)/2 & E_2 &= \alpha + \beta \\ \pi_3 &= (2z_1 + z_2 - z_3 - 2z_4 - z_5 + z_6)/\sqrt{12} & E_3 &= \alpha + \beta \\ \pi_4 &= (z_2 - z_3 + z_5 - z_6)/2 & E_4 &= \alpha - \beta \\ \pi_5 &= (2z_1 - z_2 - z_3 + 2z_4 - z_5 - z_6)/\sqrt{12} & E_5 &= \alpha - \beta \\ \pi_6 &= (z_1 - z_2 + z_3 - z_4 + z_5 - z_6)/\sqrt{6} & E_6 &= \alpha - 2\beta \end{aligned} \qquad (9.39)$$

These m.o.s are symbolically represented in Fig. 9.8. We have non-localized molecular orbitals that are again non-localizable.

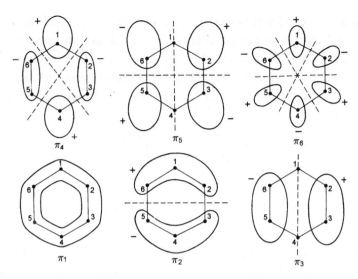

Fig. 9.8 Representation of benzene π m.o.s. Dashed lines indicate nodal planes perpendicular to the nuclear plane of the molecule.

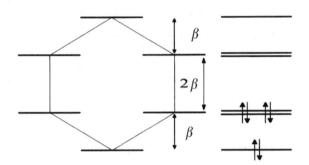

Fig. 9.9 Energy diagram for the π molecular orbitals of benzene and the regular hexagon analogue.

Figure 9.9 shows the corresponding energy diagram, the energy differences showing a parallel with the geometric features of a regular hexagon inscribed in a circle of radius 2β.

The sum of the energies of the six π bonding electrons is

$$2 \times (E_1 + E_2 + E_3) = 6\alpha + 8\beta \tag{9.40}$$

to be compared with the corresponding value for three localized π m.o.s (as in ethylene) which is $6\alpha + 6\beta$. Within the Hückel model, there is a stabilization energy of 2β associated with the occurrence of non-localizable orbitals.

The weak reactivity of benzene is in part attributed to this additional stabilization. That stabilization energy is often called the *delocalization energy* or *resonance energy* of benzene and is taken as a measure of the aromaticity of the molecule. The expression resonance energy comes from the description of the delocalized π system in valence-bond theory in terms of the resonant hybrid

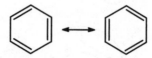

Problem 9.11   Relate the additional stabilization associated with delocalized m.o.s with the model of the particle in the box studied in Chapter 2.

There is an important pattern of the Hückel π molecular orbitals which shows up already in ethylene, butadiene and benzene and is common to many conjugated hydrocarbons. It has an essentially topological basis. The π m.o.s occur in bonding and anti-bonding pairs with energy values symmetrically arranged with respect to α: $\alpha + k$ and $\alpha - k$, respectively. Furthermore, in each pair of π m.o.s, the coefficients c_n are numerically the same but change sign for alternate a.o.s in the corresponding expressions, as for example

$$\pi_2 = (z_2 + z_3 - z_5 - z_6)/2$$
$$\pi_4 = (z_2 - z_3 + z_5 - z_6)/2 \tag{9.41}$$

in benzene.

Problem 9.12   Confirm that the pattern above is present in ethylene and butadiene.

This uniformity is generally observed for the so-called *alternate hydrocarbons* which are the conjugated hydrocarbons whose C atoms in the classical structural formula can be labelled alternately by a star, no star, a star, no star, and so on, such that no two starred or unstarred atoms are neighbours when the labelling is complete. This is the case, for example, for benzene

and for butadiene

but not for azulene

If the number of starred and unstarred C atoms is not the same, the difference corresponds to the occurrence of non-bonding π m.o.s. The π molecular orbitals for many alternate and non-alternate hydrocarbons have been tabulated (see, for example, ref. 13)

Another uniformity concerns the special stability of aromatic hydrocarbons having $4n + 2$ π electrons; $n = 1$ for benzene, $n = 2$ for naphthalene, etc. The reason lies in the observation that, in such cases, all bonding m.o.s are filled whereas the anti-bonding ones are vacant. In these hydrocarbons, except for the lowest energy π m.o. (two electrons), the bonding π m.o.s invariably occur in pairs of degenerate orbitals – $4n$ electrons; hence a total number of $4n + 2$ bonding electrons.

Problem 9.13 The energies of the π m.o.s of monocyclic hydrocarbons of n atoms can be directly obtained from the vertical coordinates of the corners of an n-sided regular polygon inscribed in a circle of radius 2β, with one corner at the lowest point (see Fig. 9.9) (ref. 121). Apply this construct to a square planar cyclobutadiene molecule (the real geometry is rectangular) and conclude that this is not an aromatic hydrocarbon.

The problem is, however, more complex than the above Hückel's rule for aromaticity suggests. In fact, there are many examples of molecules that possess the structural features required by Hückel's rules but are not aromatic (see ref. 122 and references therein).

The attention of the reader is called to some recent calculations that provide an alternative way of understanding the special stability of conjugated systems (refs. 40 and 97). For a recent paper explaining the role of the σ-skeleton, see ref. 123.

9.6 π Electron densities and bond orders

In the previous discussions, expressions for normalized π molecular orbitals have been considered under two assumptions: that the atomic orbitals used as basis are already normalized functions and that overlap integrals can be neglected. For example, for the π_1 m.o. of butadiene (see also Problem 9.10) we have

$$\int \pi_1^2 \, dv = 0.37^2 \times \int z_1^2 \, dv + 0.60^2 \times \int z_2^2 \, dv + 0.60^2$$
$$\times \int z_3^2 \, dv + 0.37^2 \times \int z_4^2 \, dv = 1. \qquad (9.42)$$

Each electron described by this m.o. is shared among the four carbon atoms according to the following weights:

0.37^2 for the atomic orbitals z_1 and z_4

0.60^2 for the atomic orbitals z_2 and z_3.

We say that the π *electron density* per carbon atom is, for that m.o. (filled with two electrons), $2 \times 0.37^2 = 0.28$ for C(1) and C(4), and $2 \times 0.60^2 = 0.72$ for C(2) and C(3). When the contribution from the other occupied m.o. π_2 is added, the result is

$$2 \times 0.37^2 + 2 \times 0.60^2 = 1 \qquad (9.43)$$

for any of the four C atoms. The four π electrons of butadiene are uniformly shared by the four C atoms.

In a general definition, the π electron density for atom i is

$$\rho_i = \sum_k n_k c_{ik}^2 \qquad (9.44)$$

where n_k is the number of electrons in the occupied m.o. k and c_{ik} is the coefficient of a.o. i in the expression of m.o. k.

Problem 9.14   Confirm that the π electron density per C atom is 1 in the benzene molecule.

It can be demonstrated that alternate hydrocarbons in their ground states have all $\rho_i = 1$ (ref. 124). Part of their special stability is attributed to this uniformity of charge distribution.

In a similar way to Eq. (9.44) we can define the π *bond order*, as a measure of the extent of π bonding for each pair of adjacent atoms, i, j:

$$p_{ij} = \sum_k^{\text{occ.}} n_k c_{ik} c_{jk}. \qquad (9.45)$$

Problem 9.15 Confirm that the π bond order in ethylene is 1 (*ignore the overlap integral*).

For butadiene, we have

$$\begin{aligned} p_{12} = p_{34} &= 2 \times 0.37 \times 0.60 + 2 \times 0.60 \times 0.37 = 0.89 \\ p_{23} &= 2 \times 0.60 \times 0.60 - 2 \times 0.37 \times 0.37 = 0.45. \end{aligned} \tag{9.46}$$

It is found that the terminal CC bonds have a slightly smaller bond order than ethylene whereas the central CC bond has almost 50% π character. These results are in qualitative agreement with the observed bond lengths.

A similar calculation for benzene leads to a π bond order of 0.67, which is higher than the classical value of 0.50 obtained on the basis of resonance between the two Kékulé structures.

Although bond lengths are also influenced by other factors, Fig. 9.10 shows a reasonable linear correlation between experimental CC bond lengths and calculated π bond orders (ref. 111).

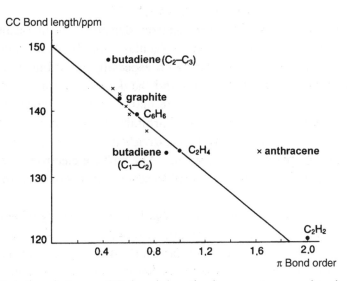

Fig. 9.10 Relation between CC bond lengths in π systems and π bond order. (Reprinted with permission from ref. 111; copyright John Wiley & Sons Ltd.)

10

Patterns in localized chemical bonds

10.1 Back to structural formulae

In Chapter 8, we reconciled the realistic visualization of molecules in terms of a global electron density distribution around an equilibrium arrangement of nuclei, as required by experiment, with the classical description of molecular structure in terms of individual atoms bonded to their neighbours. For a comprehensive analysis of the various ways of representing molecular structures, readers are invited to consult ref. 125.

In theoretical terms, the total electron density in a molecule is easily expressed in terms of the occupied molecular orbitals. Additional information is gained from the m.o. approach especially regarding the electronic energy for ground and excited states and the detailed features (e.g. phase) of individual m.o.s. Molecular orbitals are mathematical functions that can be constructed as linear combinations of orbitals of the contributing atoms, in a process where the atoms lose their individuality, except for the respective nuclei and, perhaps, the core electrons. The valence electrons are described by functions which, in general, extend to several atoms or even to the whole molecule.

As we have seen in Chapter 8, a bridge between a theory of non-localized (canonical) m.o.s and classical structures involving the sharing of electron pairs between bonded atoms can be established by appropriate unitary transformations of the canonical m.o.s. Such mathematical transformations change the basic functions used in the construction of the electronic wavefunction of the molecule but leave this wavefunction unaltered. In this way localized (or quasi-localized) functions directly associated with electron lone pairs and formal bonds can be defined. Although this definition is accomplished without altering the electronic wavefunction for the whole molecule, the physical meaning of each localized function, *per se*, has some limitations, because in general they are not independent. In particular, there is in general no direct relation between such functions and ionization energies or electron excitation; this is because they are not eigenfunctions of a hamiltonian, hence they cannot be associated with an energy. For that reason, we kept the usual designation 'localized molecular orbitals' but with the last word in inverted commas: 'orbitals'. However, for the interpretation of some other molecular properties, the minimized residual interactions between quasi-localized molecular 'orbitals' are not very important and, so, the direct use of a localized bond description is quite justified. That is the case for properties such as bond energies and electric dipole moments, as well as the features of the total electron density distribution with which those properties are directly associated.

In Chapter 9, we found that some geometrical arrangements of nuclei do not allow an equivalent set of localized molecular 'orbitals' to be defined. In such cases, there are non-localizable canonical m.o.s; these structures are presented as resonant hybrids in the classical valence-bond description.

Classical structural formulae constitute an absolutely essential component of chemical language for many reasons. In particular, they reflect the existence of patterns which facilitate the understanding of an enormous amount of data. The construction and use of mechanical or computer-generated molecular models are based on such patterns concerning bond angles and bond lengths. For example, the combined symbol O–H in any formula carries with it information regarding bond length and bond energy irrespective, to a good approximation, of the molecule being considered, be it an alcohol, an acid, or water. The length of a C–H bond in hydrocarbons is always around 108 pm, its stretching force constant around $500\,\mathrm{N\,m^{-1}}$, and its bond energy about $405\,\mathrm{kJ\,mol^{-1}}$. Similar constancy of bond parameters is found for other bonds. For example, addition of a CH_2 unit to a saturated hydrocarbon chain leads to a calculated increase in the enthalpy of atomization of $346\,\mathrm{kJ\,mol^{-1}}$ (for the new C–C bond) $+\,2\times411\,\mathrm{kJ\,mol^{-1}}$ (for the two new C–H bonds) $=1168\,\mathrm{kJ\,mol^{-1}}$, with an error of only 1 or 2 per cent when compared with experiment (ref. 40). Accordingly, the incremental increase in the standard enthalpy of formation of a normal hydrocarbon CH_3–$(CH_2)_n$–CH_3 per methylene group is $-20.6\pm1.3\,\mathrm{kJ\,mol^{-1}}$, it being possible to fit the experimental values with the following group additivity scheme

$$\Delta H_f^0(298\mathrm{K}) = 2\times42.3 + n\times20.6\ \mathrm{kJ\,mol^{-1}} \tag{10.1}$$

where $42.3\,\mathrm{kJ\,mol^{-1}}$ is the contribution of each methyl group (ref. 71).

We have previously (Chapter 8) discussed the valence-shell electron pair repulsion method as a predictive model of molecular geometry which, although not based explicitly on the orbital concept, requires prior know-ledge of a classical structure in terms of localized electron pairs. The high predictive power of this model – partially rationalized in terms of the concept of local charge concentration due to Bader – is a clear manifestation of certain patterns in molecular geometries in relation to the classical molecular structures. For example, molecules of the so-called type AX_3E, with atoms A and X connected by single bonds and E representing an electron lone pair localized in A, invariably possess a pyramidal geometry. We stress the expression 'localized in A', because if there is significant dispersal of the lone pair density over receptive groups – as, for example, in planar $N(SiH_3)_3$ with involvement of d orbitals of silicon – planar geometries may be preferred (ref. 126).

In this chapter, some additional patterns related to the localized bond description of molecular structure will be presented, especially in connection with the position of each element in the Periodic Table.

10.2 Bond energies and the Periodic Table

It is the almost constancy of XY bond energies, for each bond order, from molecule to molecule that makes the definition of average values for bond energies a particularly useful one. Table 10.1 shows such average values in relation to the Periodic Table, as taken from ref. 127.

For groups 1–14, one finds in general a decrease of the XX bond energy with the increase of atomic radius (atomic number) in each group. For example, the bond energy is larger for Li–Li ($105\,\text{kJ}\,\text{mol}^{-1}$) than for K–K ($49\,\text{kJ}\,\text{mol}^{-1}$); in the former case, atomic orbitals of principal quantum number $n = 2$ are involved, whereas it is $n = 4$ in the latter. The value for H–H, which involves 1s orbitals, is much larger ($432\,\text{kJ}\,\text{mol}^{-1}$). Similarly, in group 14 the major change takes place between C–C ($346\,\text{kJ}\,\text{mol}^{-1}$) and Si–Si ($222\,\text{kJ}\,\text{mol}^{-1}$) and this observation has a counterpart in the corresponding bond lengths: 154 pm for C–C, 234 pm for Si–Si, 244 pm for Ge–Ge, 280 pm for Sn–Sn, and 288 pm for Pb–Pb.

The general decrease of XX bond energy with increasing atomic number is also observed for groups 15–17, but only after the second row. The N–N, O–O, and F–F bond energies are abnormally small: 167, 142, and $155\,\text{kJ}\,\text{mol}^{-1}$, respectively, whereas the values for P–P, S–S, and Cl–Cl are, respectively, 220, 240, and $240\,\text{kJ}\,\text{mol}^{-1}$. This deviation from the expected trend is frequently attributed to strong repulsions between the electron lone pairs resulting from the much smaller sizes of the second row atoms than their heavier congeners. Thus, such repulsions would be much stronger for F_2,

Table 10.1 Bond energies (kJ mol^{-1}) in relation to the Periodic Table (taken from ref. 127)*

Key

A–X
Self { A–A, A=A, A≡A
Bonds to Hydrogen A–H
Oxygen A–O, A=O
Fluorine A–F

Element symbol, Electronegativity

Pauling Equation:

$$D_{A-B} = (D_{A-A}D_{B-B})^{1/2} + 96.5(\chi_A - \chi_B)^2$$

1	2	13	14	15	16	17	18
H, 2.20 432 ... 459 565							He
Li, 0.98 105 243 ... 573	Be, 1.57 208 ... 444 632	B, 2.04 293 ... 389 536 636 613	C, 2.55 346 602 835 411 358 799 485	N, 3.04 167 418 942 386 201 607 283	O, 3.44 142 494 459 142 494 190	F, 3.98 155 ... 565 ... 155	Ne
Na, 0.93 72 197 ... 477	Mg, 1.31 129 ... 377 513	Al, 1.61 ... 272 ... 583	Si, 1.90 222 318 318 452 640 565	P, 2.19 ≈220 ≈481 322 335 544 490	S, 2.58 240 425 363 ≈523 284	Cl, 3.16 240 ... 428 218 249	Ar
K, 0.82 49 180 ... 490	Ca, 1.00 105 ... 460 550	Ga, 1.81 113 ... ... ≈469	Ge, 2.01 188 272 ... ≈470	As, 2.18 146 ≈380 247 301 389 440	Se, 2.55 172 272 276 ... ≈351	Br, 2.96 190 ... 362 201 250	Kr, 3.00 50
Rb, 0.82 45 163 ... 490	Sr, 0.95 84 ... 347 553	In, 1.78 100 ... ≈523	Sn, 1.80 146 ... ≈450	Sb, 2.05 121 ≈295 ... ≈420	Te, 2.10 126 218 238 ... ≈393	I, 2.66 149 ... 295 201 278	Xe, 2.60 84
Cs, 0.79 44 176 ... 502	Ba, 0.89 44 ... 467,561 578	Tl, 2.04 ... 439	Pb, 2.33 ... ≈360	Bi, 2.02 ... 192 ... 350	Po, 2.00 ...	At, 2.20 ... 116	Rn ≈131

*The A–A values for group 2 are quite different from those tabulated in ref. 183; for example, 55 kJ mol^{-1} for Be–Be and 5 kJ mol^{-1} for Mg–Mg.

with a bond distance of 142 pm, than for Cl_2, with a bond length of 199 pm. However, this explanation needs some consideration. It is not correct to think of those repulsions as an effect to be added to the result otherwise obtained from a molecular orbital calculation. In particular, any self-consistent field (SCF) m.o. calculation already includes all the interelectronic repulsions (although in averaged terms). As clearly shown in ref. 40, the lone pair repulsions in question are of the same nature as those found for two helium atoms and reproduced by m.o. theory. Since the anti-bonding m.o. for two $1s^2$ He atoms is more anti-bonding than the bonding m.o. is bonding (see also page 79), the existence of two anti-bonding electrons and two bonding electrons in a hypothetical He_2 renders this molecule very unstable. A similar situation occurs with the two π bonding and the two π anti-bonding m.o.s of F_2, all doubly occupied, which partly offset the σ bond effect. In Cl_2, the π interactions involving 3p orbitals are much less important and the difference is much smaller between the bonding degree of the π bonding m.o.s and the anti-bonding degree of the anti-bonding π m.o.s.

A similar effect is found for example for the single bonds N–O, P–O, and As–O, with energies 201, 335 and 301 kJ mol^{-1}, respectively; and for N–F, P–F, and As–F whose bond energies are, respectively, 283, 490 and 440 kJ mol^{-1}.

Problem 10.1   Give an interpretation for the following facts:

(a) *Fluorine F_2 is very reactive.*
(b) *Many explosives include N–N and N–X bonds.*

The small size of the first row atoms shows up in high ionization potentials too and in higher electronegativities than the elements below in the same group.

With the notable exception of N–N, O–O, and F–F, the X–X bond energy in general increases in each period, thus following the increase of effective nuclear charge, for example, 105, 293 and 346 kJ mol^{-1} for Li–Li, B–B and C–C, respectively.

Since π bonding is in general less effective than σ bonding, the X=X bond energies are less than twice the corresponding X–X values, with the exception of N=N and O=O due to the anomalous low values for N–N and O–O. For example, for C–C we have 346 kJ mol^{-1} and for C=C we have 602 kJ mol^{-1}.

Problem 10.2   Justify the instability of O_3 in relation to O_2 in terms of bond energies (that is, $\Delta H < 0$ for $2O_3 \rightarrow 3O_2$).

In accordance with the prediction that π bonding becomes much less effective beyond the first row, the energy of double bonds in general decreases in each group (for groups 14–16). For example, the value for O=O is $494\,kJ\,mol^{-1}$ and that for Se=Se is $272\,kJ\,mol^{-1}$, whereas for C=C and Si=Si the respective values are 602 and $318\,kJ\,mol^{-1}$. The same applies to the triple bonds of N_2 ($942\,kJ\,mol^{-1}$) and P_2 ($481\,kJ\,mol^{-1}$).

Problem 10.3   Justify the following observations:

(a) *The most stable allotropic form of oxygen is O=O, whereas that of sulphur involves S–S bonds.*

(b) *Carbon dioxide is made of O=C=O molecules, whereas silicon dioxide is a giant covalent structure involving single Si–O bonds.*

Problem 10.4   Compare the atomization energy of P_4 *(six P–P bonds) with the energy of the* $P\equiv P$ *bond and do the same for a hypothetical* N_4 *molecule of structure identical to* P_4 *in comparison with the* $N\equiv N$ *bond energy. Comment.*

The bond energy values given in Table 10.1 correspond to ΔU_0 values, that is, change of internal energy at 0 K (see below) for

$$X_2 \rightarrow 2X \qquad (10.2)$$

or

$$XY_n \rightarrow X + nY \qquad (10.3)$$

divided by n. It should be noted that these bond energies are different from the corresponding standard enthalpy changes ΔH^{o}_{298} (see also ref. 128). First, they differ by the work of expansion associated with the above reactions:

$$\Delta U^{o}_{298} = \Delta H^{o}_{298} - P\Delta V. \qquad (10.4)$$

For example, for H_2, with $\Delta H^{o}_{298} = 436.0\,kJ\,mol^{-1}$ and $P\Delta V = RT = 2.5\,kJ\,mol^{-1}$, it is $\Delta U^{o}_{298} = 433.5\,kJ\,mol^{-1}$. On the other hand, ΔU^{o}_{298} differs from ΔU_0 because the former value includes the difference between the translational energy of 2 mol of H and the translational energy + rotational energy of 1 mol of H_2, at 298 K, which is $3RT - 5RT/2 = 1.2\,kJ\,mol^{-1}$. Hence, the value $\Delta U_0 = 432.3\,kJ\,mol^{-1}$.

The latter value includes the zero-point vibrational energy for H_2: $25.5\,kJ\,mol^{-1}$. It is therefore less than the difference between the energy of 2 mol of H and 1 mol of H_2, if these molecules were not vibrating (besides being at rotational and translational rest). For this difference we would have $(423.3 + 25.5)\,kJ\,mol^{-1} = 457.8\,kJ\,mol^{-1}$ which is the value to be compared with the calculated energies for H_2 and 2H, in their fundamental states

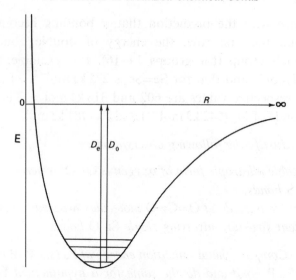

Fig. 10.1 Distinction between D_0, the dissociation energy of a diatomic molecule from the lowest vibrational level, and D_e, the binding energy, which does not take vibrational energy into account.

(D_e in Fig. 10.1). To avoid confusion, some authors distinguish between *dissociation energy* ($D_0 = \Delta U_0$) and *binding energy* (D_e), according to whether the vibrational energy is included or not when comparing X_2 with $2X$.

10.3 The octet rule and the writing of structural formulae

The concept of valence electrons was first advanced by Lewis in 1902, well before the advent of quantum mechanics and even before the nuclear structure for the atom was established by Rutherford and co-workers. A periodicity of 8 in atomic number for the properties of elements of small Z and the chemical inactivity of the noble gases led him to propose eight valence electrons for the atoms of noble gases, except helium with two, corresponding to structures of high stability. Such a stability would be associated with an 'average' distribution of eight electrons arranged according to the vertices of a cube. The stability of ions which are isoelectronic with atoms of rare gases would be rationalized in a similar way, as noted by Kossel in 1919. That is the case, for example, of O^{2-}, F^-, Na^+, Mg^{2+} and Al^{3+}, all isoelectronic with Ne.

When in 1916 Lewis published his ideas concerning chemical bonding due to pairs of electrons shared by two atoms, this *octet rule* was extended to atoms in molecules, the shared electrons being counted for each atom.

The stability of molecules would then be attained if they could adopt structures in which, through electron sharing, H atoms would have the electron configuration of helium and the remaining atoms an octet of valence electrons as in neon.

Indeed, the great majority of substances which are made of elements of low atomic number, are represented by structural formulae which are in agreement with such an octet rule (a doublet rule for H). This is, for example, the case for hydrocarbons and many other organic molecules, even when their structures are described by resonant hybrids. The rule therefore represents an extremely useful tool for the quick writing of structural formulae.

Problem 10.5 Confirm that the structural formulae for N_2, O_2, F_2, HF and CO are in accord with the octet rule.

When more than one structural formula for a given chemical species can be written in agreement with the octet rule, this usually means that resonance must be considered to describe the real structure adequately. The benzene molecule and the carbonate ion, CO_3^{2-}, are well-known examples:

There are examples of significantly different structural formulae for a given molecular formula, all obeying the octet rule, and then additional considerations are needed to decide about the correct structure. For example, the structure of ozone, O_3, is

and not

which would be a highly strained molecule.

Also, carbon dioxide has the 'symmetrical' structure

$$\overline{\underline{O}} = C = \overline{\underline{O}} \quad \longleftrightarrow \quad |\overline{\underline{O}} - C \equiv O| \quad \longleftrightarrow \quad |O \equiv C - \overline{\underline{O}}|$$

and not

$$\overline{\underline{C}} = O = \overline{\underline{O}} \quad \longleftrightarrow \quad |\overline{\underline{C}} - O \equiv O| \quad \longleftrightarrow \quad |C \equiv O - \overline{\underline{O}}|$$

In general, for second row atoms, the most electronegative atoms are located at the ends of the molecule, corresponding to a minimized repulsion of negative electric poles. Similarly, metal complexes contain a central electropositive atom surrounded by more electronegative ligands. When the electronegativities of the atoms involved are similar, then isomers may occur. One example is ClOCl and ClClO.

Problem 10.6   Contrary to Cl_2O, F_2N does not show isomerism. Justify.

An alternative way to justify the 'symmetrical' structure of CO_2 is to start with the isoelectronic species N_3^- and the respective distribution of ionic charge: -0.67 at the terminal N atoms and $+0.33$ at the central atom, according to simple m.o. theory. Thus, the new nuclear charges $+8$ and $+6$ for carbon dioxide are expected to lead to greater stabilization with an arrangement $+8, +6, +8$ than with $+6, +8, +8$. For the same reason, N_2O has the geometry NNO instead of the 'symmetrical' structure NON (see also ref. 40).

Problem 10.7   Interpret the asymmetrical geometry of N_2O:

(a) *in terms of the hypothetical addition of a proton to one of the N nuclei of N_3^-;*

(b) *in terms of the hypothetical rearrangement of the protons of the nuclei of CO_2 in order to obtain N_2O.*

Problem 10.8   Considering that the meta carbon positions of pyridine are negatively charged, whereas the ortho and para positions are positively charged, predict which structures of diazine (diazabenzene) and triazine (triazabenzene) are more stable.

It is also possible, using the octet rule and the concept of resonant hybrids, to interpret and predict non-uniform electron densities in some π systems. For example, π electron populations higher than 1 in the *ortho* and *para* positions of the aniline molecule can be rationalized in terms of resonance between the following structures

The less than unity π electron populations in the same positions of nitroben-zene require resonance involving structures that deviate from the octet rule:

In the above scheme, the arrows correspond to the differences in the structures to be considered in the description of nitrobenzene as a resonant hybrid.

Notwithstanding the pragmatic interest of the octet rule, it should be noted that it provides, by itself, no explanation for the stability of molecules but an interesting and useful *correlation* with the stability of the noble gases. Even for those monoatomic ions which are isoelectronic with atoms of noble gases, such a correlation already asks for an *interpretation* and then an *explanation*. The interpretation lies in the high ionization energies of atoms of rare gases and isoelectronic ions as well as in their low electron affinities. The explanation can be found in the corresponding fundamental electron configurations: high effective nuclear charges justifying high ionization energies, and complete valence electron shells justifying low electron affinities (additional electron in a high energy level). It is in this light that the usual reference to a special stability of atoms having a complete valence shell should be understood. In this way, it is possible to avoid the common idea conveyed in teaching that open-shell atoms are less stable because of some (almost affective) aversion to having an incomplete shell.

We have previously shown that classical structural formulae can be reproduced on the basis of molecular orbital theory. The stability associated with molecular structures obeying the octet rule results from the fact that all bonding (and non-bonding) molecular orbitals are occupied.

A shared doublet in the case of H, hence single bonds involving H atoms, is a result of H contributing with just one valence orbital. Similarly, the reference to an octet is a consequence of other atoms contributing with four valence orbitals. If the number of valence orbitals for C, for example, was three and not four, then we would be talking of a sextet rule.

Problem 10.9   *Assume that the nitrogen atom, with five valence electrons, had three valence orbitals.*

(a) *How many valence molecular orbitals should be considered for* N_2?
(b) *What would be the bond order if three bonding m.o. and two anti-bonding m.o.s were filled?*
(c) *What would be the structural formula of* N_2?

It is stressed that the octet rule is certainly a pattern of second row atoms, when only four valence atomic orbitals per atom can participate in bonding. If d orbitals are also contributing, then an 18-electron rule is expected corresponding to the 18 valence electrons of the next 'inert' gas in the Periodic Table. However, a proper m.o. determination is required in order to ascertain the role of anti-bonding m.o.s. We will come back to this problem in Chapter 11 when extending the orbital concept to metal complexes.

The octet rule also holds in many cases involving third row atoms, for example in HCl and H_2S, where the atoms only use four valence atomic orbitals in the construction of the m.o.s. In cases such as the so-called hypervalent molecules PCl_5 and SF_6, it has been common practice to consider, respectively, 10 and 12 electrons surrounding the central atom and regard these examples as exceptions to the octet rule, explained by some involvement of d orbitals of the central atom:

Similarly, the following structural formulae are frequently encountered

(resonance with another three, all equivalent)

(resonance with another five, all equivalent)

$$\overset{\textstyle |\underline{O}|}{\underset{\textstyle |\underline{O}-H}{\overset{\textstyle \|}{\underline{O}=S-\underline{O}-H}}}$$

$$\overset{\textstyle |\underline{O}|}{\underset{\textstyle |\underline{O}|}{\overset{\textstyle \|}{|\overline{O}-Cl-\underline{O}|}}}{}^{-} \qquad \text{(resonance with another three, all equivalent)}$$

all bond orders being correlated with bond lengths (see refs. 129 and 130). For example, the SF distance in SF_6 (156 pm) is almost equal to that in SF_2 (159 pm). If the experimental bond length 143 pm for SO_2 and SO_3 is assigned to a double S=O bond, then the values 142 pm and 157 pm, respectively, for the peripheral SO bonds and for the inner SO(H) bonds in H_2SO_4 point to two S=O bonds and two S–O bonds, as shown above. In agreement with those findings, the value 149 pm for SO_4^{2-} is attributed to a bond order of 1.5. Similarly, the values 139, 143, 149, 158 and 167 pm in ClO_2^+, ClO_4^-, ClO_3^-, ClO_2^- and ClO^-, respectively, point to the following bond orders: 2, 1.75 (see structure above), 1.67, 1.5 and 1.

However, it is well-known that bond energies and bond lengths depend not only on the corresponding bond order but also on its polarity. Thus, recent calculations (refs. 131 and 132) indicate that the participation of 3d orbitals has been overemphasized and that octet structures, when associated with the polarity of the bonds, constitute a better description (see also refs. 133 and 134). For example:

$$\overset{\textstyle |\overline{O}|}{\underset{\textstyle |\underline{O}|}{|\overline{O}-S-\underline{O}|}}{}^{2-} \quad \longleftrightarrow \quad \overset{\textstyle |\underline{O}|}{\underset{\textstyle |\underline{O}|}{|\overline{O}-S-\underline{O}|}}{}^{2-}$$

$$\overset{\textstyle |\underline{O}|}{\overset{\textstyle \|}{|\overline{O}-S-\underline{O}|}} \qquad \text{(resonance with another two, all equivalent)}$$

$$|\overline{O}=S-\underline{O}| \quad \longleftrightarrow \quad |\overline{O}-S=\underline{O}|$$

and analogously for the chlorate and phosphate ions.

For SF_6, molecular orbital calculations involving not only the sulphur 3s and 3p orbitals but also the 3d orbitals (ref. 40) lead to approximately the same results as without the 3d orbitals. This ought not to be surprising in view of the great difference of energy between these orbitals and the fluorine valence orbitals. Thus, because of symmetry, we have essentially four doubly occupied bonding m.o.s and two ligand-located non-bonding lone pairs. The bond order of each SF linkage is 2/3 and a resonance hybrid must be considered involving structures like:

$$
\begin{array}{c}
\mathrm{F}^{-} \\
\mathrm{F}^{-} \\
\mathrm{F}\!-\!\overset{2+}{\mathrm{S}}\!-\!\mathrm{F} \\
\underset{\mathrm{F}}{\big|}\;\;\mathrm{F}
\end{array}
$$

Finally, a word about molecules such as BF_3 for which the following structure is usually written

$$
\begin{array}{c}
|\overline{\mathrm{F}}| \\
| \\
\mathrm{B} \\
\diagup \quad \diagdown \\
|\overline{\mathrm{F}}| \quad\;\; |\overline{\mathrm{F}}|
\end{array}
$$

with less than an octet around the central atom and in accordance with the Lewis acid character of this compound. However, a resonance hybrid involving structures of the type

$$
\begin{array}{c}
|\mathrm{F}| \\
\| \\
\mathrm{B} \\
\diagup \quad \diagdown \\
|\overline{\mathrm{F}}| \quad\;\; |\overline{\mathrm{F}}|
\end{array}
$$

has also been proposed in accordance with a bond length of 131 pm which is smaller than the value for the single bonds in BF_4^- (139 pm) (ref. 135).

10.4 The conservation of the sum of bond orders

The bond orders we are considering are the ones defined in terms of the classical structural formulae, which can be related to the numbers of bonding and anti-bonding electrons. They are not the bond orders derived from molecular orbital numerical calculations as given by Eq. (9.45). An additional uniformity involving such bond orders and chemical equations can be found, directly related to the octet rule. For example, by considering the following chemical equations

$$3H_2 + N_2 \rightarrow 2NH_3$$

$$2C_2H_6 + 7O_2 \rightarrow 4CO_2 + 6H_2O$$

$$3C_2H_2 \rightarrow C_6H_6 \tag{10.5}$$

$$CO_3^{2-} + 2H_3O^+ \rightarrow CO_2 + 3H_2O$$

$$CH_3CO_2H + H_2O \rightarrow CH_3CO_2^- + H_3O^+$$

and by writing the structural formulae of all species involved, it is found that the sum of the bond orders is conserved when going from reactants to products: 6, 28, 15, 10, 10, respectively. The same observation applies to reactions involving condensed phases such as

$$C(c) + O_2(g) \rightarrow CO_2(g) \tag{10.6}$$

providing the unit that repeats itself in the crystalline structure of C is considered, namely

$$\begin{array}{c} C/4 \\ | \\ C \\ \diagup \; | \; \diagdown \\ C/4 \quad | \quad C/4 \\ C/4 \end{array}$$

for diamond, with $1 + 4 \times 1/4 = 2$ atoms C for four single bonds. In this case, the sum of the bond orders is 4.

In cases like

$$H + Cl \rightarrow HCl$$

$$PCl_5 \rightarrow PCl_3 + Cl_2$$

$$CF_2 + F_2 \rightarrow CF_4 \tag{10.7}$$

$$CO_3^{2-} + 2H^+ \rightarrow CO_2 + H_2O$$

the rule no longer holds. The difference with respect to the first group of examples is that, now, some species have structures that do not conform with the octet rule (or the doublet rule for H). As we shall see below, this is not accidental.

In any structure that obeys the octet rule, the atoms other than H contribute with four valence a.o.s and with at least four valence electrons, the total number of valence electrons in the molecule being even. In such cases, all bonding and non-bonding m.o.s are filled. Thus, for each molecule i, the sum of the bond orders is half the difference between the number of bonding electrons and the number of anti-bonding electrons

$$(\text{BO})^i = (N_b^i - N_{ab}^i)/2. \tag{10.8}$$

This expression can be written alternatively as follows, involving the number of non-bonding electrons if present,

$$(\text{BO})^i = [N_b^i - (N^i - N_b^i - N_{nb}^i)]/2 = N_b^i - (N^i - N_{nb}^i)/2. \tag{10.9}$$

Now, since all n^i bonding and non-bonding m.o.s are filled, we have

$$\begin{aligned}
(\text{BO})^i &= 2n_{bmo}^i - N^i/2 + n_{nbmo}^i \\
&= n_{bmo}^i + n_{abmo}^i + n_{nbmo}^i - N^i/2 \\
&= n_{mo}^i - N^i/2 \\
&= n_{ao}^i - N^i/2
\end{aligned} \tag{10.10}$$

where n_{mo}^i is the number of valence m.o.s of molecule i and n_{ao}^i is the number (the same) of the corresponding a.o.s. Since there is conservation of the total number of valence electrons and of the total number of valence atomic orbitals contributed by the various atoms (conservation of the number of atoms of each element), when going from reactants to products, then the sum of the bond orders is also conserved.

For the conditions considered in the discussion above – octet rule obeyed for each species, with filled bonding and non-bonding m.o.s – it is found that the effective net number of bonding electrons is constant: $N_b^i - N_{ab}^i$. This is the very number of electrons represented by the strokes between atomic symbols in the classical structural formulae, in terms of which the bond orders are defined here. For a more detailed discussion see ref. 136.

11

The concept of molecular orbitals in other systems

The molecular orbital model as a linear combination of atomic orbitals introduced in Chapter 4 was extended in Chapter 6 to diatomic molecules and in Chapter 7 to small polyatomic molecules where advantage was taken of symmetry considerations. At the end of Chapter 7, a brief outline was presented of how to proceed quantitatively to apply the theory to any molecule, based on the variational principle and the solution of a secular determinant. In Chapter 9, this basic procedure was applied to molecules whose geometries allow their classification as conjugated π systems. We now proceed to three additional types of systems, briefly developing firm qualitative or semiquantitative conclusions, once more strongly related to geometric considerations. They are the recently discovered spheroidal carbon cluster molecule, C_{60} (ref. 137), the octahedral complexes of transition metals, and the broad class of metals and semi-metals.

11.1 The C_{60} molecule

Buckminsterfullerenes are, by now, well-known allotropes of carbon, especially since they were at the centre of the 1996 Nobel Prize in Chemistry attributed to the British chemist Harold Kroto (1939–) and to the American chemists Robert Curl (1933–) and Richard Smalley (1943–). The C_{60} molecule is the best representative of this new class of stable clusters of carbon having the shape of a football:

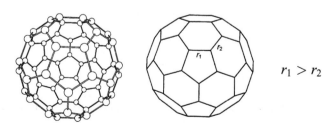

$$r_1 > r_2$$

There are 60 corners in the above polyhedron (truncated icosahedron) corresponding to the nuclei of the C atoms. There are $60 \times 4 = 240$ valence electrons. Each C atom is bonded to three other carbon atoms, with the total number of covalent bonds being 90. In graphite, each C atom is also trigonally bonded to other carbon atoms, but unlike C_{60}, the geometric arrangement is planar and a $\sigma-\pi$ separation similar to ethylene or benzene is possible. Such a separation is not rigorously possible in C_{60}. One can, however, still consider 90 σ-type bonds approximately described by quasi-localized σ molecular 'orbitals' involving $90 \times 2 = 180$ bonding electrons, together with a set of 30 inter-pentagonal π-type CC bonds. If sp^2 carbon orbitals are considered, bent σ m.o.s are constructed and the radially directed p orbitals are canted with respect to each other by about $23°$. A three-dimensional π system thus arises on which a Hückel calculation can be performed. Due to the high symmetry of this system, several sets of degenerate m.o.s are obtained, as shown in Fig. 11.1.

A closed-shell molecule is predicted, in agreement with experiment. The π m.o.s are non-localizable and a delocalization energy per carbon atom is calculated that is larger (about 60%) than that of benzene and comparable to that of graphite (ref. 138). The highest occupied m.o.s (HOMO) constitute a fivefold degenerate set, whereas the lowest unoccupied m.o.s (LUMO) are triply degenerate, a significant energy gap existing between both sets. Early extended Hückel calculations involving all valence atomic orbitals (ref. 139) and *ab initio* Hartree–Fock calculations confirmed these

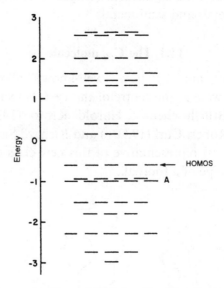

Fig. 11.1 Hückel π molecular orbitals for C_{60} (see, for example, ref. 138).

results (ref. 140). Geometry optimization SCF studies (refs. 140–142) predicted two distinct bond lengths: longer values for the five-membered rings (longer than the benzene value), and shorter values for those bonds which are common to neighbouring six-membered rings (shorter than the benzene value), in agreement with experiment.

The almost spherical shape of the C$_{60}$ molecule favours an approximate approach to the 60 π electrons, based on an extension of the free particle in the box model to a spherical surface which we have considered in Chapter 2 (page 39). The various quantized energy levels related to an angular momentum quantum number L are given by expression (2.73)

$$E_L = L(L+1)h^2/8\pi^2 I \tag{11.1}$$

where

$$I = mR^2 \tag{11.2}$$

is the inertia moment of the particle of mass m on a surface of radius R. In the present case, $R = 350$ pm.

For each L value, $2(2L+1)$ electrons can be allocated: 2 for $L=0$, 6 for $L=1$, 10 for $L=2$, 14 for $L=3$, 18 for $L=4$ and the remaining 10 for $L=5$. The latter energy level would thus be only partly filled and the lowest energy absorption transition (selection rule $\Delta L = 1$) would involve an electron promotion from $L=4$ to $L=5$. The calculated wavelength from this model is 398 nm, which is in surprisingly good agreement with the experimental value, 404 nm (ref. 143).

Problem 11.1   Confirm the value 398 nm for the transition $L=4$ to $L=5$.

However, there is an important deviation from experiment. Such a model predicts that the ground state of C$_{60}$ should be paramagnetic

h	$L=5$	↿ ↿ ↿ ↿ ↿ ↿ ↿ ↿ ↿ —
g	$L=4$	⇅ ⇅ ⇅ ⇅ ⇅ ⇅ ⇅ ⇅ ⇅
f	$L=3$	⇅ ⇅ ⇅ ⇅ ⇅ ⇅ ⇅
d	$L=2$	⇅ ⇅ ⇅ ⇅ ⇅
p	$L=1$	⇅ ⇅ ⇅
s	$L=0$	⇅

which is not the case. This problem is removed when due account is taken of the fact that the geometry is not rigorously spherical but that of a truncated icosahedron. Then, all levels with $L>2$ are split (ref. 144) and the energy level diagram becomes:

	— — —
	— —	
$L = 5$	⥮ ⥮ ⥮ ⥮ ⥮
	⥮ ⥮ ⥮ ⥮
$L = 4$	⥮ ⥮ ⥮ ⥮ ⥮
	⥮ ⥮ ⥮
$L = 3$	⥮ ⥮ ⥮ ⥮
$L = 2$	⥮ ⥮ ⥮ ⥮ ⥮
$L = 1$	⥮ ⥮ ⥮
$L = 0$	⥮

This model qualitatively reproduces the Hückel m.o. diagram of Fig. 11.1 for the lowest and highest occupied energy levels.

More recently, SCF calculations have extended to endohedral complexes of C_{60} with various ions (see, for example, ref. 145), and density-functional theory (Kohn–Sham orbitals) has been applied (see, for example, ref. 146).

11.2 Octahedral complexes of transition metals

A complete study of the molecular orbitals for an octahedral complex such as $[Cr(CN)_6]^{4-}$ or $[Co(NH_3)_6]^{3+}$ would require linear combinations of all the valence atomic orbitals of the metal and of the ligands. An approximation is to take the metal valence a.o.s (nine a.o.s for a metal of the first transition series (five 3d orbitals, one 4s and three 4p orbitals)) together with six a.o.s from the ligands, one for each atom directly bonded to the metal atom. In general, these six a.o.s are quasi-localized molecular 'orbitals' (see Chapter 8), which point from the ligand to the metal and have essentially non-bonding character:

They can be, for example, a carbon hybrid orbital, n, of the cyanide ion

$$|C\equiv N|^-$$

or one of the valence electron pair orbitals of F^-, Cl^-, and O^{2-}, or the lone pair N orbital of NH_3, or one of the lone pair O orbitals of OH^- or H_2O, etc.

There will be, thus, 12 valence electrons from the ligands to be considered plus N electrons coming from the metal ion. The number of a.o.s to be combined will be $6+9=15$ leading to the construction of 15 m.o.s which are confined to the central metal atom and the six bonded atoms from the ligands. This does not mean, however, that they are localized in each formal metal–ligand bond. In order to obtain localized molecular 'orbitals', a unitary transformation of the Slater determinant which is built on the basis of the occupied canonical m.o.s would be needed. The metal participates in these localized functions with six d^2sp^3 hybrid orbitals. Once again, we use inverted commas in the expression localized molecular 'orbitals' to stress that no energy can be associated with these functions. Thus, it cannot be inferred from such a description that the electrons of the metal–ligand bonds have the same energy. Those 'orbitals' are not independent, in the sense that there are residual interactions between them that cannot be ignored for purposes related to electronic energy. In particular, there are interactions (resonance integrals in simple m.o. theory) between the hybrid orbitals themselves and between each ligand n orbital and the various metal hybrid orbitals other than that pointing to it.

The traditional description of the complex formation in terms of single dative covalent bonds between the ligand and the metal is consistent with the proposal of six bonding m.o.s, which are doubly occupied (12 electrons), plus six anti-bonding m.o.s leaving three non-bonding ones. This hypothesis is supported by the existence of three d orbitals which are orthogonal to the ligand n orbitals, as is shown further below. The non-bonding orbitals will be occupied with a number of electrons equal to the N valence electrons contributed by the metal:

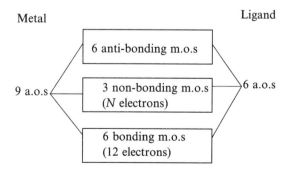

As with other symmetrical systems (Chapters 7 and 9), the form of the m.o.s can easily be found by establishing intermediate linear combinations of the n orbitals contributed by the ligands. These are symmetry-adapted linear combinations that can be constructed using group theory (see, for example, ref. 40):

$$n_1 + n_2 + n_3 + n_4 + n_5 + n_6 \qquad \text{non-degenerate combination: } a_{1g}$$

$$\left.\begin{array}{c} n_1 - n_2 \\ n_3 - n_4 \\ n_5 - n_6 \end{array}\right\} \quad \text{triply degenerate combination: } t_{1u}$$

$$\left.\begin{array}{c} n_2 - n_3 + n_5 - n_6 \\ 2n_1 - n_2 - n_3 + 2n_4 - n_5 - n_6 \end{array}\right\} \quad \text{doubly degenerate combinations: } e_g.$$

$$(11.3)$$

The various combinations are named according to the degree of degeneracy and to the g ('gerade', i.e. even) or u ('ungerade', i.e. odd) parity (see page 75): the a_{1g} and the two e_g combinations have an even number of nodes, whereas each of the three t_{1u} combinations have just one node.

Problem 11.2   Confirm that the symmetry-adapted combinations above are orthogonal.

This method reduces the number of associations to be considered with the a.o.s of the central atom, because some of them correspond to zero overlap. The metal a.o.s are, for the first transition series, the 4s orbital, the three 4p orbitals and the five 3d orbitals. The latter are expressed below in polar coordinates (see page 46), together with the appropriate connection to cartesian coordinates. These expressions are real linear combinations of the complex forms of the d functions which correspond to each of the possible values of the magnetic quantum number m_ℓ. The corresponding angular parts are depicted in Fig. 11.2.

$$\psi_{3d_{z^2}} = \frac{1}{81\sqrt{6\pi}} \left(\frac{Z}{a_0}\right)^{3/2} \rho^2 e^{-\rho/3} (3\cos^2\theta - 1) \propto e^{-\rho/3}(z^2 - r^2)$$

$$\psi_{3d_{x^2-y^2}} = \frac{1}{81\sqrt{2\pi}} \left(\frac{Z}{a_0}\right)^{3/2} \rho^2 e^{-\rho/3} \sin^2\theta \cdot \cos 2\phi \propto e^{-\rho/3}(x^2 - y^2)$$

$$\psi_{3d_{xy}} = \frac{1}{81\sqrt{2\pi}} \left(\frac{Z}{a_0}\right)^{3/2} \rho^2 e^{-\rho/3} \sin^2\theta \cdot \sin 2\phi \propto e^{-\rho/3}xy$$

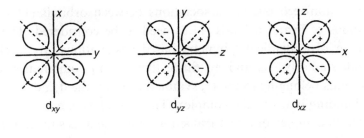

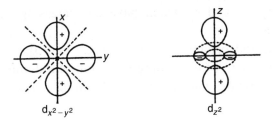

Fig. 11.2 Graphical representation of the angular part of the five 3d orbitals.

$$\psi_{3d_{xz}} = \frac{\sqrt{2}}{81\sqrt{\pi}} \left(\frac{Z}{a_0}\right)^{3/2} \rho^2 e^{-\rho/3} \sin\theta \cdot \cos\theta \cdot \cos\phi \propto e^{-\rho/3} xz$$

$$\psi_{3d_{yz}} = \frac{\sqrt{2}}{81\sqrt{\pi}} \left(\frac{Z}{a_0}\right)^{3/2} \rho^2 e^{-\rho/3} \sin\theta \cdot \cos\theta \cdot \sin\phi \propto e^{-\rho/3} yz$$

Although the five d orbitals of an isolated atom are isoenergetic, they can be divided into two groups according to the orientation of the corresponding axes with respect to the x, y and z cartesian axes: orbitals d_{xy}, d_{yz} and d_{xz} make a 't' group, whereas d_{z^2} and $d_{x^2-y^2}$ form an 'e' group.

We can, thus, recognize the following groups of orbitals contributed by the metal atom:

s orbital	a_{1g}
p_x, p_y, p_z orbitals	t_{1u}
d_{xy}, d_{yz}, d_{xz} orbitals	t_{2g}
$d_{z^2}, d_{x^2-y^2}$ orbitals	e_g.

(11.4)

Problem 11.3

(a) *Confirm the classification of the d_{z^2} and $d_{x^2-y^2}$ orbitals as functions of g parity.*

(b) *Justify the difference between t_{1u} and t_{2g} above.*

Only the non-zero overlap associations between orbitals (11.4) and the combinations of ligand orbitals (11.3) need to be considered. They are the ones that involve identical symmetry-based designations: a_{1g} in (11.3) and (11.4) and, similarly, t_{1u} and e_g. Since there is no combination of the n_i ligand orbitals belonging to the t_{2g} symmetry group, the d_{xy}, d_{yz}, d_{xz} orbitals are non-bonding m.o.s in the complex. The typical m.o. energy diagram for an octahedral complex of a first transition series metal is shown in Fig. 11.3. The m.o. of lowest energy (a_{1g}) has no nodes, that is, it is bonding everywhere. It is followed by t_{1u} and e_g which have an increasing number of nodes (of angular origin).

Usually, the ligand orbitals have less energy than the 3d orbitals of the metal. As a consequence, the bonding m.o.s reflect a dominant contribution from the ligand orbitals, whereas the anti-bonding m.o.s are dominated by

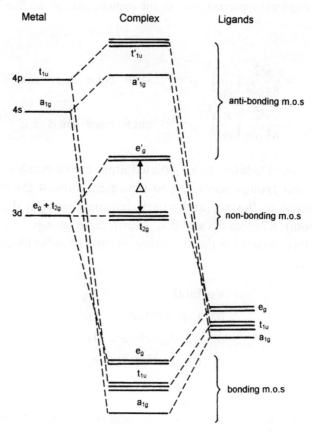

Fig. 11.3 Typical m.o. energy diagram for an octahedral complex of a metal of the first transition series.

the contributions from the metal orbitals. For example, the main contribution to the first anti-bonding m.o., e'_g, comes from the d_{z^2}, $d_{x^2-y^2}$ orbitals (e_g). Since, in general, the more stable the bonding m.o.s, the higher the energy of the anti-bonding m.o.s, the difference $\Delta = E_{e'_g} - E_{t_{2g}}$ is usually taken as a measure of the strength of the metal–ligand bonds.

This model based on a shortened version of m.o. theory is known as *ligand field theory*, a name reminiscent of the so-called *crystal field theory* that approaches the problem by considering the effect of six negative point charges, octahedrally placed around a metal atom, on the energies of the various d orbitals:

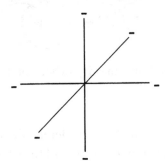

In the latter model, electrostatic repulsion between the point charges and the d electrons is intuitively expected to be stronger for those orbitals having greater density along the cartesian axes, that is d_{z^2} and $d_{x^2-y^2}$. This is supported by a full electrostatic calculation involving all the point charges (not just one for each d orbital) (ref. 40). As a consequence, a splitting of the d orbitals in two groups is proposed, with the e_g group lying higher in energy. In spite of the crudity of the model, this conclusion is in good agreement with the more elaborate ligand field method and, thus, sustains similar comparisons with the spectroscopic, magnetic and thermodynamic experimental data that can be directly related to the t_{2g} and e'_g levels.

A d orbital splitting is also obtained by the so-called angular orbital model, which is the m.o. analogue of crystal field theory. The energy of interaction of the six ligand n orbitals with the d orbitals (d_{z^2} and $d_{x^2-y^2}$) is related to the corresponding overlap integral which is a function of the angular behaviour of the n orbitals with respect to the cartesian axes (ref. 40).

The separation $\Delta = E_{e'_g} - E_{t_{2g}}$ is called the *ligand field splitting parameter*. As we shall see in Chapter 12, this parameter is related to the energy of the electronic transitions responsible for the colour of many complexes, because the splitting is usually small enough to involve transitions by absorption of visible radiation.

The m.o. energy diagram of Fig. 11.3 is modified if the ligand is also involved in π-type interactions with the metal. This is a common situation with most ligands, not only those having valence π orbitals, such as CN^-, but also ligands such as Cl^-, OH^-, etc. Figure 11.4 illustrates the non-zero overlap between a d orbital of group t_{2g} with a ligand orbital having π symmetry relative to the metal–ligand bond axis.

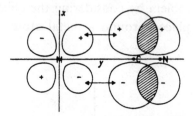

Fig. 11.4 π Interaction involving the d_{xy} orbital of the metal and one of the π m.o.s of cyanide ion as a ligand.

Two extreme situations can be considered: ligands that are *Lewis bases* (for example, Cl^-, I^-, OH^-) or *Lewis acids* (for example CO). In both cases, there are filled orbitals of π symmetry, but in the latter there are also low-lying vacant anti-bonding π m.o.s. There will be electron donation from the ligand to the metal in the case of Lewis bases, but the opposite for Lewis acids. The alterations in that part of the m.o. diagram corresponding to the ligand field splitting are qualitatively shown in Fig. 11.5. The result is a decrease of Δ for Lewis bases and an increase for Lewis acids as ligands: the t_{2g} orbitals acquire, respectively, anti-bonding and bonding character.

With the limitations of the 'aufbau' principle, already encountered in the discussion of the filling of 3d and 4s atomic orbitals for transition metal atoms (Chapter 5), these diagrams can be used for establishing the electron configurations of the complexes. The critical question is the distribution of a number of electrons equal to N, the number of valence electrons contributed by the metal, among the t_{2g} and e'_g levels of the complex. If the ligand field splitting is large (*strong field* case), the electrons are assigned preferably to the non-bonding t_{2g} orbitals. If the ligand field splitting is small (*weak field* complexes), then states of lower energy can be obtained by also using the anti-bonding m.o.s e'_g providing those states have higher spin multiplicity. This is a manifestation of Hund's rules (Chapter 5).

The complex $[Cr(CN)_6]^{4-}$ is an example of a strong field complex, whereas $[Cr(H_2O)_6]^{2+}$ is a weak field case. There are $N=4$ electrons to be attributed corresponding to a sub-configuration d^4 of Cr^{2+}. The (lowest

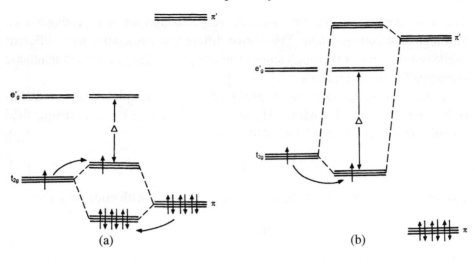

Fig. 11.5 Changes in the ligand field splitting Δ due to interaction of the non-bonding t_{2g} orbitals with ligand orbitals of π symmetry: ligands acting as Lewis bases (a) or Lewis acids (b).

energy) sub-configuration for the hexacyanochromate(II) ion is

$$(t_{2g})^4$$

whereas for the hexaquochromium(II) ion it is

$$(t_{2g})^3, (e_g')^1.$$

In the first case, the lowest energy level assumes the existence of two unpaired spins,

whereas, in the latter, all the electrons have the same spin:

The two configurations are named, respectively, *low-spin configuration* and *high-spin configuration*. The above differences associated with different orbital contributions (through spin–orbit coupling) lead to different magnetic moments for the complexes.

A more dramatic example is provided by the complexes of cobalt(III) (sub-configuration $3d^6$) with NH_3 and F^-. $[Co(NH_3)_6]^{3+}$ is a strong field complex (large Δ value) and its configuration is

$$(t_{2g})^6$$

whereas $[CoF_6]^{3-}$ is a weak field one (small Δ value) with configuration

$$(t_{2g})^4, (e'_g)^2.$$

Accordingly, the hexaminocobalt(III) ion is diamagnetic

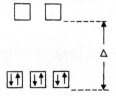

whereas the hexafluorcobaltate(III) ion is paramagnetic:

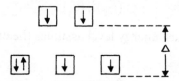

Problem 11.4   Establish the high-spin and the low-spin configurations for the octahedral complexes of Fe(III) and Co(II).

The above hexacoordinated complexes of Co(III) illustrate the so-called 18 electron rule, according to which special stability is attained for 18 valence electrons around the central atom corresponding to the next group 18 'inert' gas. The number 18 is appropriate to a full occupancy of all bonding and non-bonding levels. However, this is not the case for the weak field $[CoF_6]^{3-}$ complex, where the low-lying occupied e'_g m.o.s do actually contribute to stabilization, by allowing higher spin multiplicity, in spite of their anti-bonding nature.

Problem 11.5   Explain why $[Ni(H_2O)_6]^{2+}$ is a stable complex in spite of having 20 and not 18 valence electrons around the metal atom.

The occupancy of the degenerate e_g' m.o.s in $[CoF_6]^{3-}$ is the basis of a distortion of the octahedral geometry: there is, in fact, an identical increase of the lengths of two opposite Co–F bonds relative to the perfect octahedral arrangement. This is a manifestation of the known Jahn–Teller theorem according to which degenerate electronic states in non-linear molecules are unstable with respect to a geometrical distortion that lifts that degeneracy.

The 18 electron rule has a parallel with the octet rule for elements of smaller atomic number. We have already commented upon the structure of SF_6 (page 240) where only apparently the octet rule is violated. The m.o.

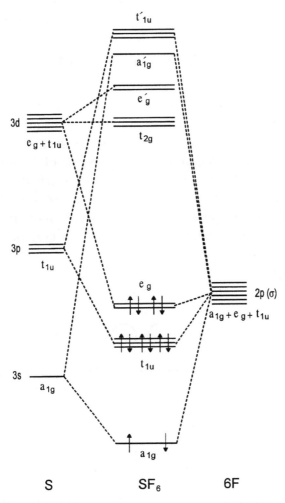

Fig. 11.6 Molecular orbital diagram for SF_6 with the inclusion of the 3d orbitals on the S atom (a small effect in view of the higher energy of the 3d orbitals relative to 3s and 3p).

diagram for octahedral SF_6, with the inclusion of the sulphur 3d orbitals (Fig. 11.6) differs from that of Fig. 11.4 for transition metal complexes because the t_{2g} and e'_g m.o.s lie higher in energy as a consequence of the higher energy of the 3d orbitals of S relative to 3s and 3p. Figure 11.6 shows that the bonding in SF_6 is essentially due to eight electrons, the four e_g electrons being essentially non-bonding. That is, there is an average S–F bond order of 2/3 (see also ref. 133).

   The example of SF_6 in this context of metal complexes serves to recall that main group molecules and transition metal complexes are not so different after all. Some of the fruitful bridges between both types of systems are provided by the so-called isolobal relationships developed by R. Hoffmann (for a discussion of this theme, see, for example, ref. 147).

11.3 The band theory of solids

Molecular orbital theory can be applied to aggregates of a virtually infinite number of atoms, as in metals or diamond, for example. Whereas in the study of metal complexes one can restrict oneself to the m.o.s directly related to the metal–ligand interactions, now it is necessary to consider the whole sample of solid. Thus, if each atom of an aggregate of N atoms contributes with one valence orbital, s, there will be N orbitals which, in principle, extend to the whole aggregate. One could still call them molecular orbitals, which is equivalent to considering the whole sample as a giant molecule.

   Among such a huge number of molecular orbitals, one is maximally bonding having no nodes:

$$\psi_1 = s_1 + s_2 + s_3 + s_4 + \cdots \tag{11.5}$$

and another is the most anti-bonding, having the maximum number of nodes between atoms

$$\psi_N = s_1 - s_2 + s_3 - s_4 + \cdots \tag{11.6}$$

Between those two extremes (corresponding to extreme values of energy) there will lie $N - 2$ m.o.s with a varying number of nodes and various energies. The result is a *band* of m.o.s having a nearly continuous band of energy values.

Problem 11.6   *Give the expressions for the three m.o.s formed from three s orbitals in a row: one bonding, one anti-bonding and one non-bonding.*

Problem 11.7   *Interpret the following form for the Hückel secular determinant for a row of N atoms contributing just one s orbital each:*

$$
\begin{vmatrix}
\alpha - E & \beta & 0 & 0 & 0 & \cdots & 0 \\
\beta & \alpha - E & \beta & 0 & 0 & \cdots & 0 \\
0 & \beta & \alpha - E & \beta & 0 & \cdots & 0 \\
\vdots & \vdots & \vdots & \vdots & \vdots & \vdots & \\
0 & 0 & 0 & 0 & 0 & \cdots & \alpha - E
\end{vmatrix} = 0
$$

The extension to three dimensions requires a knowledge of the geometric arrangement of the atoms. For a square prism arrangement of 16 atoms (ref. 148)

one can begin by making intermediate linear combinations for each side edge involving four s orbitals. Thus, by adding and subtracting $s_1 + s_2$ and $s_3 + s_4$ and by adding and subtracting $s_1 - s_2$ (or $s_2 - s_1$) and $s_3 - s_4$ (or $s_4 - s_3$), we get for one edge:

$$
\begin{array}{ccccc}
s_1 & + & - & + & - \\
s_2 & + & + & + & + \\
s_3 & + & + & - & - \\
s_4 & + & - & - & +
\end{array}
\tag{11.7}
$$

There are four associations of the type (11.7) which can now be combined in pairs corresponding to each face:

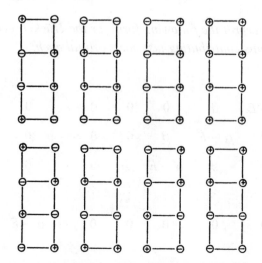

This procedure is finally extended to the three dimensions (schemes reproduced with permisssion from ref. 148):

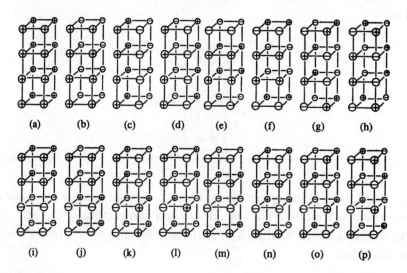

By counting the number of nodes for each one of the 16 orthogonal linear combinations, we get the total number of bonding and anti-bonding interactions as shown in Table 11.1.

From Table 11.1 a semi-quantitative energy diagram can be constructed. Thus, the bonding levels of increased energy are: a, i, (b, c, e), (j, k, m) (Fig. 11.7). The number of m.o.s per unity energy change is not uniform. A *density of states* can be defined showing a maximum in the centre of the band.

Table 11.1 *Number of bonding and anti-bonding interactions for the orthogonal linear combinations of 16 s orbitals having a square prismatic arrangement*

Combination	Bonding interactions	Anti-bonding interactions
a	28	0
b	20	8
c	20	8
d	12	16
e	20	8
f	12	16
g	12	16
h	4	24
i	24	4
j	16	12
k	16	12
l	8	20
m	16	12
n	8	20
o	8	20
p	0	28

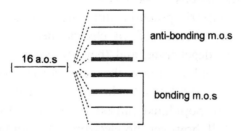

Fig. 11.7 Molecular orbital energy diagram for a square prism arrangement of 16 atoms, each one contributing a valence s orbital.

For N atoms having only one valence orbital (s orbital), the $N/2$ low-lying m.o.s will be filled if the atom configuration is s^1, but the whole band would be complete if the configuration was pure s^2. When both s and p orbitals are available for bond formation, an s band and a p band will be formed. The top of the s band may lie below or above the bottom of the p band. That is, there may be a *band gap* between the highest energy m.o. of the s band and the lowest energy m.o. of the p band. However, there will always be a difference in angular momentum between s and p levels.

In the case of metals where the bonds are weak and the separation of their s and p orbitals is small, a virtually continuous band of levels occurs. Open-shell configurations mean that only part of the whole band will be occupied:

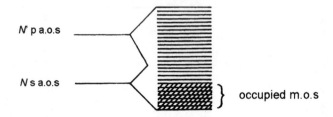

The fact that there is almost a continuum of energy values means that the electron distribution will be very sensitive to temperature. For $T=0\,\mathrm{K}$, the highest energy level occupied is called the *Fermi level*. An incompletely filled band is termed a *conduction band* because the highest energy electrons can adjust rapidly to an applied electric field and give rise to electrical conductivity and other properties characteristic of metals. This means that an entering electron finds a low energy level in which to move.

It is the presence of empty electronic levels immediately above the Fermi level that enables the metals to be *conductors*. The perturbation brought about by an applied electric potential difference can be interpreted in the following way. In the absence of an electric field, electron states with opposite momenta are degenerate, and they are equally populated; but, in the presence of a potential difference, the states corresponding to motion in one direction are stabilized relative to those travelling in the opposite direction, and become more populated (unless the band is full). A net current arises. If the band is full, however, no electron redistribution is possible and all states remain equally populated. No net current flows and the material is an *insulator*.

An increase of temperature of a metal, besides exciting the valence electrons, also intensifies the vibrations of the atoms: the solid lattice. The result is a dispersion of the electron motion and the electric conductivity decreases.

Metals are aggregates of atoms having weakly bound valence electrons. The same applies to the π electrons of graphite, which is also a conductor. Diamond, however, is a giant structure of strong covalent bonds where all electrons are tightly bound. Now, there is a band gap between occupied bonding m.o.s and vacant anti-bonding m.o.s of about $500\,\mathrm{kJ\,mol^{-1}}$. Since each C atom contributes four valence atomic orbitals and four valence electrons, for N atoms there will be $2N$ bonding m.o.s which are completely filled and $2N$ vacant anti-bonding m.o.s (at 0 K):

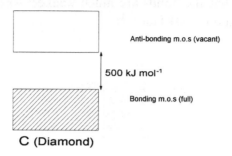

Anti-bonding m.o.s (vacant)

500 kJ mol^{-1}

Bonding m.o.s (full)

C (Diamond)

The vacant levels are too high in energy to allow adjustment of the electron populations by an electric field, and diamond is an insulator. The full sub-band is called a *valence band*.

The canonical m.o.s of diamond are delocalized over the entire crystal. However, as we have seen in Chapter 8 for other systems, the occupied m.o.s can be the object of a unitary transformation leading to a set of equivalent and quasi-localized molecular 'orbitals'. This is why the structure of diamond can (for some purposes) be described in terms of the overlap of sp^3 hybrid orbitals, four for each C atom. As we have seen in Chapter 8, we must stress that such an alternative description cannot be used to infer information about electron energies. In particular, the localized bond description of the structure of diamond does not imply that all valence electrons have the same energy. This would be the case only if the sp^3–sp^3 bonds were independent. It is because of residual interactions such as β' and β''

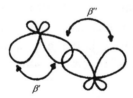

that bonding and anti-bonding bands occur.

Problem 11.8 From the expressions $(1/2)(s+x+y+z)$ *and* $(1/2)(s-x+y-z)$ *for two* sp^3 *hybrid orbitals of C, find that* β' *is given by* $\beta' = (1/4)(\alpha_s - \alpha_p)$ *where* $\alpha_s = \langle s|\hat{H}|s\rangle$ *and* $\alpha_p = \langle p|\hat{H}|p\rangle$.

At high temperatures, a few electrons will be excited to the anti-bonding levels (conduction band) and some redistribution of electron populations due to applied electric fields becomes possible. The electric conductivity increases with temperature. This effect is more marked in silicon and can also be brought about by interaction with light. Crystalline silicon has a structure similar to diamond but the bonds are much weaker. As a consequence, the band gap drops to about $100 \, \text{kJ mol}^{-1}$:

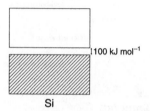

Silicon possesses the properties of a *semiconductor*. Electric thermoconductivity and photoconductivity are two features of semi-metals.

An increase of conductivity of a semiconductor can also be obtained by replacement of some atoms (to an extent of about 1 in 10^9) by dopant atoms, that is, atoms of another element, which can either increase or decrease the number of electrons in the material. In the first case, the conduction band becomes partially populated; in the latter, the dopant introduces holes in the previously full valence band, empty levels becoming accessible in this way.

Band theory is also applicable to solid compounds. Thus, in sodium chloride we can define two sets of bands, one formed mainly from the 3s orbitals of the Na atoms and the other from the 3s and the three 3p orbitals of the Cl atoms. In accordance with the greater electronegativity of Cl, the chlorine s and p bands (four) lie below the s band of Na and are fully occupied by all the valence electrons (eight for each NaCl pair). The band gap between this valence band and the empty s band of Na is almost $700 \, \text{kJ mol}^{-1}$ and the material is an insulator. Again, we can transform this delocalization picture into a localized one, with the limitations already stressed. Thus, four full chlorine bands are equivalent to an array of individual chloride ions, whereas an empty sodium band is equivalent to a collection of Na^+ ions.

12

Orbitals in action

Orbitals are a mathematical concept and are not real entities. However, they provide the basis for the interpretation of both the structure and changes of chemical substances. Indeed, besides the more fundamental concepts of energy and probability briefly discussed in Chapter 1, the orbital concept – which in fact accommodates the concepts of energy and probability in an intimate way – is, perhaps, the concept with the strongest impact in modern Chemistry. In this chapter, an illustration is given of the direct relationships between orbitals and chemical reactivity and between orbitals and spectroscopy. In the latter, excited states are an implicit part of the problem. However, no detailed treatment of the important relation between excited states and reactivity will be performed in this book.

12.1 Orbitals and chemical reactivity

This discussion will be mainly qualitative. We will begin with a classic example of chemical reactions directly related to the electron density distribution: the electrophilic aromatic substitution reactions of the benzene derivatives aniline and phenol:

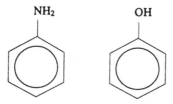

The substituent groups $-NH_2$ and $-OH$ interact with the π orbital system of the benzene ring leading to an increase of π electron density at the *ortho* and *para* positions:

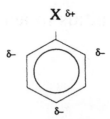

As a consequence, electrophilic aromatic substitution reactions occur more easily than in benzene, and with preference for attack at the *ortho* and *para* positions.

It is noted that the correlation above is with the π electron distribution and not with the overall electron density at each C atom. Indeed, the alternation of π population is accompanied by an opposing polarization of the σ density distribution. For aniline, the calculated changes in the π and σ contributions to the charge on each C relative to benzene are (ref. 91):

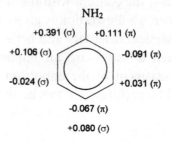

Problem 12.1 *On the basis of π electron density, interpret the observation that a nitro group in a benzene ring deactivates the ring towards electrophilic substitution.*

Several specific reactions can be easily interpreted by involving the so-called *frontier orbitals*, that is, the occupied molecular orbital of highest energy and the vacant molecular orbital of lowest energy. They are usually named HOMO and LUMO, respectively, for 'highest occupied molecular orbital' and 'lowest unoccupied molecular orbital'. The importance of this approximate approach was first stressed by the Japanese chemist Kenichi Fukui (1918–1998, 1981 Nobel laureate in Chemistry) (see ref. 149).

Simple examples are found among some redox reactions (ref. 150). For example, F_2 is reduced to F^- (and not to F_2^{2-}) by gain of two electrons, because these electrons must be assigned to the LUMO of F_2 which is an anti-bonding molecular orbital. Similarly, HBrO is reduced to Br^- by a two-electron process

$$2e^- + H^+ + HBrO \rightarrow H_2O + Br^- \tag{12.1}$$

and not to Br_2 by a one-electron process

$$e^- + H^+ + HBrO \rightarrow H_2O + (1/2)Br_2 \tag{12.2}$$

because the relevant LUMO is anti-bonding in the region of the Br–O bond. The two electrons accepted by HBrO lead to the breaking of that bond

$$H-Br-O \xrightarrow{2e^-} H-O^- + Br^- \tag{12.3}$$

whereas one electron would only weaken the Br–O bond in the radical formed (unpaired electron mainly centred on Br). Moreover, an expected reaction between radicals would increase the population of the anti-bonding LUMO, leading, again, to the breaking of the Br–O bond with formation of Br^-.

For those reactions in which a given species acts as an electron donor, it is the corresponding HOMO that is particularly important. In contrast, it is mainly the LUMO that must be considered in the case of an electron acceptor molecule. An approximate approach ignores alterations in the remaining m.o.s and bases the discussion upon the symmetry of the orbitals.

A good example is provided by the bimolecular nucleophilic substitution (S_N2), associated with the Walden inversion, the attack by a base B occurring from the side opposite to the atom to be replaced:

The nucleophile B transfers electrons from its HOMO to the LUMO of the substrate. The latter is the three-node function shown in Fig. 12.1 together with the B lone pair for two orientations of attack. An attack perpendicular to the C–Cl bond axis would lead to an unfavourable activated complex, because for such a mutual orientation the total overlap between the lone pair orbital of B and the LUMO of CH_3Cl is small. This is not the case for orientation 1 in Fig. 12.1. With this interaction, the LUMO becomes partially

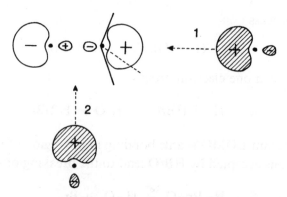

Fig. 12.1 Relevant m.o.s (LUMO of CH_3Cl and HOMO of a base *B*), for two orientations of attack, in a S_N2 reaction with Walden inversion.

populated

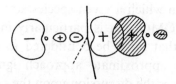

and this weakens the C–Cl bond, the region where the LUMO is anti-bonding. In addition, it is argued that the negative overlap between the C–*B* bond being formed and the 1s orbitals of the H atoms favours the migration of these atoms to the opposite side (see, for example, ref. 13):

$$H-\overset{\displaystyle H}{\underset{\displaystyle H}{C}}-B$$

Addition to a carbonyl group is another example of the application of frontier orbitals. In accordance with the electronegativity difference, the π bonding m.o. of a $>C=O$ group has a larger contribution from the oxygen 2p a.o., the opposite occurring for the π' anti-bonding m.o., which is the LUMO of the group. Hence, nucleophiles (electron donors) bond to the C atom, whereas electrophiles attack the O atom. In density-functional theory (see page 164), the points of attack are found from the contour plots of the so-called Fukui function $f(\mathbf{r})$, which measures how sensitive a system's electronic chemical potential is to an external perturbation at a particular point (Fig. 12.2).

Yet another application of the frontier orbitals lies in the field of the relative stability of isomers and conformers. Let us consider propene,

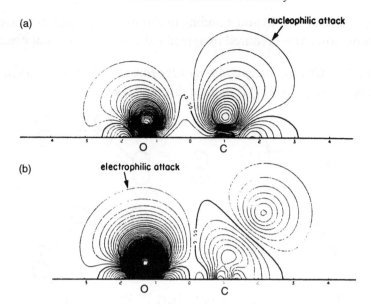

Fig. 12.2 Fukui contour maps for H_2CO: (a) $f^+(\mathbf{r}) \approx \rho(\mathbf{r})$ (LUMO) contours on the plane perpendicular to the nuclear plane; (b) $f^-(\mathbf{r}) \approx \rho(\mathbf{r})$ (HOMO) contours on the nuclear plane. (Adapted with permission from ref. 77; copyright 1989, Oxford University Press.)

CH_3–CH=CH_2. For any of the conformations (a) and (b)

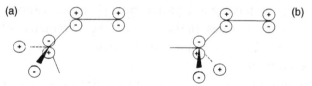

the HOMO has a π bonding character in the C=C bond, an anti-bonding effect in the C–C bond region (radial node) and a bonding character in the methyl group. It is noted that the π symmetry extends to the $-CH_3$ group although there is no formal π bonding with this group. This is an example of *hyperconjugation*. In conformer (a), one C–H axis (of $-CH_3$) and the CC axis of C=C are eclipsed, whereas they have a staggered orientation in conformer (b):

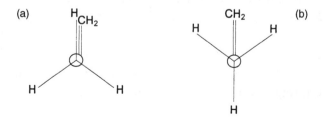

Therefore, the end-to-end anti-bonding in this m.o. is greatest for case (a) and so conformation (b) is favoured in agreement with experimental observation.

Problem 12.2   Give an interpretation of the fact that the trans *configuration of 1,3-butadiene (see page 217) is more stable than* cis:

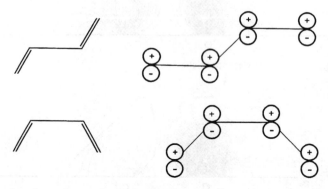

In 1965, the American chemists Robert B. Woodward (1917–1979, 1965 Nobel laureate in Chemistry) and Roald Hoffmann (1937–, 1981 Nobel laureate in Chemistry) (ref. 151), well aware of the stereospecificity of many unimolecular cyclization reactions of open conjugated molecules, e.g. *cis*-1,3-butadiene, developed a vast application of frontier orbitals to this and similar types of reactions.

The intramolecular cyclization reactions and the cycloaddition reactions are characterized by having a negative entropy of activation. The so-called Woodward–Hoffmann rules make use of the symmetry of the frontier orbitals (*the principle of orbital symmetry conservation*) to enable a qualitative interpretation of the occurrence and stereospecificity of those reactions. According to this model, the formation of the necessary σ bond at the expense of π orbitals occurs through overlap of orbital lobes of the same sign: bonding character in the region of the incipient σ bond. But the way this overlap is attained depends on the number of π electrons involved and on the conditions in which the reaction is carried out. Thus, if the cyclization is carried out by heating (thermally controlled reaction), the reactant (only vibrationally excited) is virtually all in the ground electronic state, and it is a pair of π electrons of the HOMO that intervene in the new σ bond. However, if the reaction is carried out photochemically, the molecule is in its first excited state π^1, π'^1, and the formation of the σ bond involves one electron in the HOMO (π) and one electron in the LUMO (π'). Because the symmetry of the LUMO is different from the HOMO, the products resulting from closure of the reactant are stereochemically different.

Let us consider, for example, the conversion of *cis,trans*-1,4-dimethyl-butadiene

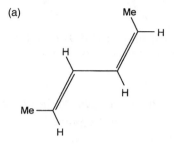

into 3,4-dimethylcyclobutene

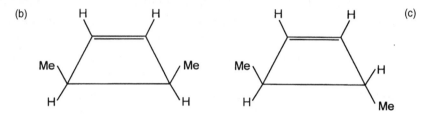

The central part of Fig. 12.3 schematically pictures the HOMO of 1,3-butadiene, already considered in a *cis* configuration.

The formation of the C(1)–C(4) σ bond via the participation of two π electrons of the HOMO requires that the $2p_\pi$ orbitals in atoms C(1) and C(4) undergo rotation in the same sense (*conrotatory mode*) so that positive (in-phase) overlap occurs (bonding effect). If they rotate in opposite directions (*disrotatory mode*) then there will be negative overlap and an anti-bonding effect. Accordingly, the conrotatory closure of (a) in a thermally controlled reaction leads to the *cis* isomer of dimethylcyclobutene (b) and not to (c).

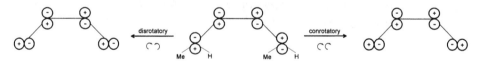

Fig. 12.3 The highest occupied m.o. (HOMO) of 1,4-dimethylbutadiene (*cis* configuration) and its participation in cyclization in the conrotatory case.

If the reaction is photochemically controlled, the excited butadiene now has an electron in a π m.o. that was empty in the ground state. This m.o. was the LUMO of ground state butadiene depicted in the central part of Fig. 12.4. Now, a concerted closure (a net bonding interaction) can only be obtained by the disrotatory mode. As a result, the *trans* isomer (c) and not (b) is the product of cyclization.

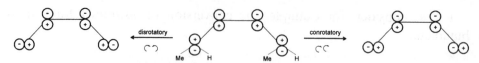

Fig. 12.4 Lowest unoccupied m.o. (LUMO) of 1,4-dimethylbutadiene (*cis* configuration) and its participation in cyclization in the disrotatory case.

A similar reasoning can be applied to hexatriene to be converted into
cyclohexadiene. The only significant change is that the number of π electrons
becomes six corresponding to six conjugated C atoms, and so there is one
more node in the frontier orbitals. Accordingly, the symmetries of the HOMO
and LUMO at the terminal C atoms are different from those for butadiene,
and cyclization occurs by the disrotatory mode or by the conrotatory mode,
respectively, for a thermally or a photochemically controlled reaction. The
general rule is that the thermal cyclization reactions of a k π-electron system
will be conrotatory for $k = 4q$ and disrotatory for $k = 4q + 2$ $(q = 0, 1, 2, \ldots)$.
For photochemical cyclizations these relationships are reversed.

*Problem 12.3 By considering the two frontier orbitals for two ethylene
molecules explain the following:*
(a) *There is no interaction between the HOMO of one molecule and the LUMO
of the other.*

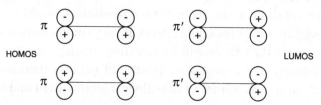

(b) *Cyclobutane is formed by dimerization of ethylene, if one molecule in a pair
is photochemically excited: π^1, π'^1.*

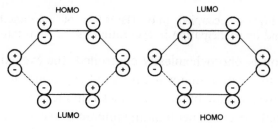

Fig. 12.5 The HOMO (butadiene)–LUMO (ethylene) and HOMO (ethylene)–
LUMO (butadiene) interactions that favour a Diels–Alder reaction.

The reaction considered in the previous problem is an example of an inter-molecular cycloaddition. Another example is the Diels–Alder reaction between a diene and an alkene. Figure 12.5 shows the favourable interactions between the HOMO of butadiene and the LUMO of ethylene, and between the HOMO of the latter and the LUMO of the former.

12.2 Orbitals and spectroscopy

In addition to the indirect relations between orbitals and spectroscopy, by means of the wavefunction of the system, there are direct connections that can be established. The very concept of energy quantization applied to particles of matter, intimately related to the orbital concept, is deeply rooted in spectroscopy: in particular the emission spectrum of atomic hydrogen.

Energy of electronic transitions

The electronic transitions in atoms and molecules (and in solids) are associated with two electronic states of the system by the Bohr frequency condition

$$\Delta E = E_2 - E_1 = h\nu. \tag{12.4}$$

They are often taken directly as transitions between atomic or molecular orbitals (or band levels, in solids). This is an approximation because it ignores changes in the remaining contributions to the total electronic energy. In particular, it assumes that the remaining orbitals are not affected in the process.

An example is the transition of one electron between an occupied π m.o. and an empty π' m.o.: a $\pi \to \pi'$ transition. Table 12.1 shows the energy difference between the lowest unoccupied m.o. (LUMO) and the highest occupied m.o. (HOMO) of polyenes, in the Hückel approximation (Chapter 9).

The expected frequency of the first (lowest energy) absorption band in the electronic spectra of these compounds should be linearly related to the ΔE

Table 12.1 *Hückel energies for the LUMO and HOMO of polyenes*

	E(HOMO)	E(LUMO)	ΔE
Ethylene	$\alpha + \beta$	$\alpha - \beta$	2β
1,3-Butadiene	$\alpha + 0.62\beta$	$\alpha - 0.62\beta$	1.24β
1,3,5-Hexatriene	$\alpha + 0.45\beta$	$\alpha - 0.45\beta$	0.90β
1,3,5,7-Octatetraene	$\alpha + 0.35\beta$	$\alpha - 0.35\beta$	0.70β

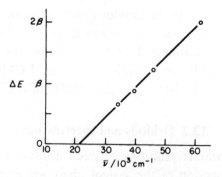

Fig. 12.6 Plot of the ΔE values of Table 12.1 against the wave numbers of the first absorption band in the electronic spectra (reprinted with permission from ref. 111; copyright John Wiley & Sons Ltd).

values. This prediction is supported by the plot in Fig. 12.6, although, due to the underlying approximations, the straight line does not pass through the origin (ref. 111). From the slope of the line, the following value for the resonance integral is obtained: $\beta = -248\,\text{kJ}\,\text{mol}^{-1}$.

Fig. 12.7 Structure of β-carotene.

Another conclusion from Table 12.1 is that conjugation of π bonds is accompanied by a steady decrease of the energy of the first $\pi \to \pi'$ transition. In long polyenes, for example carotenes (Fig. 12.7), that transition falls into the visible region of the spectrum and the compound has colour (absorption band around 451 nm).

Problem 12.4   Give a justification for the change of colour of phenolphthalein with pH:

pH 1-8 (colourless)   pH 8-10 (pink)

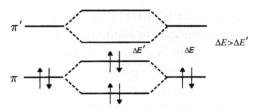

Fig. 12.8 $\pi \rightarrow \pi'$ Transition energy decreases due to conjugation of two double bonds.

If a molecule contains double bonds that are well separated from each other, then $\pi \rightarrow \pi'$ transitions occur independently by absorbing UV photons of wavelength 190 nm. But if the double bonds are adjacent, the overlap of their π orbitals must be taken into account and a new energy diagram applies where the separation LUMO–HOMO decreases. Figure 12.8 gives a simple m.o. diagram for two conjugated π (and π') orbitals.

Problem 12.5   Interpret the decrease of the energy of the first $\pi \rightarrow \pi'$ transition in polyenes with the increasing length of the conjugated chain on the basis of the model of a particle in a unidimensional box (page 31).

Problem 12.6   The π excitation energy increases on going from 4,4'-dimethyl-biphenyl and 3,3'-dimethylbiphenyl to 2,2'-dimethylbiphenyl. Explain.

Besides the $\pi \rightarrow \pi'$ transitions, the $n \rightarrow \pi'$ transitions involving non-bonding orbitals n and anti-bonding π' m.o.s are of particular importance, although they usually lead to weak bands (as we shall see below, these transitions are, in principle, forbidden). In the carbonyl group, for example, the two types of transitions occur. The non-bonding orbital can be either σ or π. Usually, the $n \rightarrow \pi'$ transitions have lower energy than the $\pi \rightarrow \pi'$ ones, for the same species. It has been shown (ref. 150) that, in some cases, a simple inspection of the Lewis formulae allows a prediction of which transition is of lowest energy. Let us take, for example, the Lewis structure for N_3^-:

$$\overline{N}=N=\overline{N}^{\,-}$$

The structure of N_3^+ (two electrons less) is

$$\overline{N}=N=\underline{N}^+ \leftrightarrow \underline{N}=N=\overline{N}^+$$

whereas that of N_3^{3-} (two electrons more) is

$$\overline{N}=\overline{N}-\overline{N}|^{3-} \leftrightarrow |\overline{N}-\overline{N}=\overline{N}^{\,3-}$$

These formulae suggest that the HOMO of N_3^- (from where two electrons are removed to form N_3^+) is essentially non-bonding, whereas the LUMO (where two electrons are added to form N_3^{3-}) is an anti-bonding π' m.o. Hence, the electronic transition of lowest energy is $n(\pi) \rightarrow \pi'$.

The $n \rightarrow \pi'$ transitions are especially sensitive to hydrogen bonds involving the non-bonding orbital n. For example, when going from a solution of acetone in hexane to a solution in chloroform, the $n \rightarrow \pi'$ transition of acetone (at about 290 nm in an inert solvent) undergoes a shift to higher frequency due to stabilization of the n orbital by hydrogen bonding to chloroform. A similar effect is observed on going from a non-polar to a polar solvent, due to a stronger stabilization of the ground state by the polar solvent because it is more polar than the excited state.

Whereas the π transitions in carotene involve delocalized m.o.s, the $n \rightarrow \pi'$ transition of acetone is localized almost entirely in the carbonyl group. The carbonyl group is an example of a *chromofore*, that is a group of atoms which give rise to characteristic excitation energies.

Problem 12.7 *By considering the residual interaction of the non-bonding 'orbitals' of pyridazine, n_1 and n_2, quasi-localized in the N atoms,*

interpret the occurrence of an $n \rightarrow \pi'$ transition at higher wavelength as compared to pyridine.

Problem 12.8 *What shift can be predicted in the $n \rightarrow \pi'$ absorption band when going from formaldehyde to benzaldehyde?*

For the octahedral complexes of transition metals studied in Chapter 11, important transitions are those of energy Δ (the ligand field splitting parameter) between the non-bonding t_{2g} orbitals and the anti-bonding e'_g orbitals. The simplest example is provided by the hexaquotitanium(III) ion, $[Ti(H_2O)_6]^{3+}$, whose fundamental valence configuration is $(t_{2g})^1$. The absorption band at 493 nm, responsible for the purple colour of the complex, is assigned to the $t_{2g} \rightarrow e'_g$ transition (Fig. 12.9).

We saw in Chapter 11 that the ligand field splitting Δ depends on the ligand. Strong field ligands lead to a greater Δ value than weak field ligands. The following sequence of ligands, essentially independent of the central

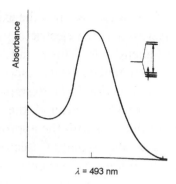

Fig. 12.9 Absorption band of $[Ti(H_2O)_6]^{3+}$ and the associated $t_{2g} \rightarrow e'_g$ transition.

metal ion, corresponds to the increasing order of Δ

$$I^- < Br^- < S^{2-} < SCN^- < Cl^- < NO_3^- < F^- < C_2O_4^{2-} < H_2O$$
$$< NCS^- < NH_3 < NH_2(CH_2)_2NH_2 < \text{bipyridine} < NO_2^-$$
$$< P(C_6H_5)_3 < CN^- < CO. \tag{12.5}$$

On account of the effects of the ligand field splitting on the absorption electronic spectra, this sequence is known as the *spectrochemical series*. The Δ splittings, being the difference between the energy of anti-bonding and non-bonding m.o.s, are a measure of the efficiency of the metal–ligand bonding: in general, the more the bonding m.o.s are bonding the more the anti-bonding are anti-bonding, hence the greater is Δ. In turn, the efficiency of the metal–ligand bonds depends on the relative energies and the overlap of the intervening orbitals, including the contribution of π interactions (see Chapter 11).

The Δ values also vary with the identity of the central metal ion. In general, the ligand field splitting increases with oxidation number and with atomic number in the same group. Particularly, the more diffuse 4d and 5d orbitals provide larger overlap with the ligand orbitals than the 3d orbitals. The following is the spectrochemical series for the metal ions:

$$Mn^{2+} < V^{2+} < Co^{2+} < Fe^{2+} < Ni^{2+} < Fe^{3+} < Co^{3+} < Mn^{4+} < Mo^{3+}$$
$$< Rh^{3+} < Ru^{3+} < Pd^{4+} < Ir^{4+} < Pt^{4+}. \tag{12.6}$$

Problem 12.9 For which of the complexes, $[Co(NH_3)_6]^{3+}$ and $[CoF_6]^{4-}$, is the ligand field splitting expected to be larger?

Probability of electronic transitions

The orbital concept is useful for the interpretation of electronic spectra in two ways: the transition energy and the transition probability. Any transition induced by electromagnetic radiation requires the existence of an interaction mechanism between the system and the electric or the magnetic component of the oscillating electromagnetic field. This requires, in turn, the system to create oscillating electric or magnetic fields, especially oscillating dipolar fields, which warrant that interaction. For example, if a molecule has an electric dipole moment, its rotations around the centre of mass can be excited by the absorption of photons with the appropriate energy. Similarly, if the molecular vibrations create oscillating electric fields, the interaction with radiation of adequate frequency becomes possible. The equivalent condition for electronic transitions is easily met, because the electrons of any system generate, in their complex motions, a variety of fluctuating electric fields.

The electronic transitions in the absorption mode can be interpreted in terms of a perturbation of the ground state wavefunction by the oscillating electric field of the radiation (ultraviolet or visible) and such a perturbation can be expressed as a mixture of the ground state wavefunction with excited state wavefunctions. The extent of mixing reflects the transition probability. The most intense transitions are those accompanied by a transient electric dipole, that is, one that arises from the changing electron distribution. For example, a transition between a 1s and a 2s orbital, which are both spherically symmetrical, corresponds to a symmetrical redistribution of charge and there is no net dipolar redistribution in such a process. Hence, the probability of this transition being brought about by a dipolar mechanism is zero. However, there is a transient dipolar redistribution of charge in a transition between an s orbital and a p orbital. The distortion of an s orbital by the dipolar electric field of the radiation can be described by a mixture with p functions but not with s functions.

Similar considerations apply to electronic transitions in molecules. Thus, a $\pi \rightarrow \pi'$ transition in a carbonyl group may be accompanied by a displacement of electron charge from the O atom to the C atom, and hence provides an interaction mechanism with the electromagnetic radiation. On the other hand, an $n(\sigma) \rightarrow \pi'$ transition is essentially a rotational migration of charge, with no dipolar effect.

The probability of a transition $i \rightarrow j$ is given by

$$P_{i \rightarrow j} = B_{ji} \rho(\nu_{ji}) \tag{12.7}$$

where $\rho(\nu_{ji})$ is the energy density of the radiation at the transition frequency ν_{ji} and where

$$B_{ji} = (d_{ji})^2 / 6\varepsilon_0 \hbar^2 \tag{12.8}$$

is one of the so-called Einstein coefficients which is related to the *transition dipole moment*:

$$d_{ji} = \langle \psi_j | -e\mathbf{r} | \psi_i \rangle \tag{12.9}$$

The operator $-e\mathbf{r}$ is the electric dipole moment operator.

Problem 12.10   By expressing $\mathbf{r}$ as $x\mathbf{i} + y\mathbf{j} + z\mathbf{k}$, show that the transition $1s \rightarrow 2s$ for the hydrogen atom has zero probability (note that x can be positive or negative).

For the hydrogen atom, the integral (12.9) is zero, unless

$$\Delta l = \pm 1 \tag{12.10}$$

for the orbitals ψ_j and ψ_i. This is the *selection rule* already encountered in Chapter 3 and based on the principle of conservation of angular momentum associated to the fact that the photon has spin (spin quantum number 1). When a photon is absorbed by an s electron, say, that electron acquires the photon angular momentum and a transition to a p orbital occurs; however, an s electron cannot be transferred to a d orbital by a single photon because the photon does not supply enough angular momentum.

Problem 12.11   The octahedral complexes of transition metals show, in general, weak colours, in comparison with the permanganate ion, for instance. In the first case, the relevant transition is $t_{2g} \rightarrow e'_g$, whereas in the latter it is a charge transfer transition between the metal and the oxygen atoms. Justify the different transition probabilities.

Spectral parameters in NMR spectroscopy

Besides the direct relations between orbitals and spectroscopy outlined above, there are many indirect relations which have to do with the interpretation of various spectral parameters in other branches of spectroscopy. We shall illustrate this with the main spectral parameters in nuclear magnetic resonance spectroscopy: NMR chemical shifts and nuclear spin–spin coupling constants.

NMR chemical shifts

The proton chemical shifts are dominated, in diamagnetic systems, by the electron population ρ_{1s} of the 1s orbital. The magnetic shielding constant for a ^{1}H nucleus is mainly expressed in terms of the so-called local diamagnetic term:

$$(\sigma_{HH})^d = 21\rho_{1s} \text{ ppm}. \tag{12.11}$$

This partly explains the δ_H values of the methane derivatives CH_3X (δ_H ppm is the chemical shift relative to the reference tetramethylsilane, $Si(CH_3)_4$; the greater the shielding constant σ_H the smaller δ_H):

	CH_4	CH_3I	CH_3Br	CH_3Cl	CH_3F
δ_H	0.23	2.16	2.68	3.05	4.26
Pauling electronegativity of X	2.20(H)	2.55	2.96	3.16	3.98.

The greater the electronegativity of X the smaller the electron density ρ_{1s} at the H atoms, hence the smaller $(\sigma_{HH})^d$.

Analogously, the ^1H chemical shifts at the *para* positions (where other effects are of minor importance) in benzene derivatives, relative to the parent hydrocarbon, are approximately given by

$$\Delta\delta_H \cong -10\Delta\rho_\pi \tag{12.12}$$

where $\Delta\rho_\pi$ is the difference of the π electron population in the *para* C atom relative to benzene (unit π electron population in each C atom). For example, there is a positive chemical shift $\Delta\delta_H = 0.38$ for the *para* position of nitrobenzene, and a negative shift of $\Delta\delta_H = -0.65$ for the same position of aniline, in accordance with the π electron distributions in the rings:

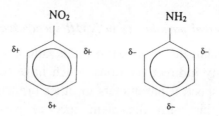

Problem 12.12 Explain the negative value $\Delta\delta_H = -0.45$ for the para position of phenol.

Considering the ^{13}C nucleus (and, similarly, most nuclei other than H), the main contribution to the magnetic shielding constant is the so-called local paramagnetic term:

$$(\sigma_{CC})^P = -\langle\varphi_{2p}|r^{-3}|\varphi_{2p}\rangle f(c_n, \varepsilon_j). \tag{12.13}$$

This term depends on the average value of r^{-3} for a 2p electron (r is the distance to the nucleus) and on the features (atomic coefficients c_n and energies ε_j) of the molecular orbitals in which the 2p orbitals of the C atom in question intervene. The average value of r^{-3} is a function of the electron population of the C orbitals. An excess of electrons leads to an expansion of the 2p orbitals, hence to a smaller absolute value of $(\sigma_{CC})^P$. As a consequence, the shielding constant σ_C increases. The expression equivalent to Eq. (12.12) for the *para* position of benzene derivatives is

$$\Delta\delta_C \cong -160\Delta\rho_\pi. \tag{12.14}$$

Problem 12.13 Calculate an approximate value for ρ_π in the para position of nitrobenzene, from the value $\Delta\delta_C = 5.8$.

In paramagnetic complexes, such as some octahedral complexes of transition metal ions, the metal–ligand interactions are responsible for transfer of electron spin from the metal to the ligands. The existence of net electron spin population in the hydrogen 1s orbitals of the ligands causes very large chemical shifts, either for high or low frequency. Three of the several mechanisms involve π interactions as follows.

(a) Transfer of electrons with unpaired spin between the non-bonding orbitals t_{2g} and the empty π' m.o.s of the ligand, leading to a net α spin density in the ligand:

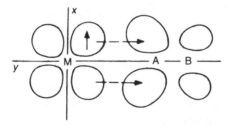

(b) Transfer of electrons of β spin from a completely filled π m.o. of the ligand into a half-filled t_{2g} orbital of the metal, leading to a net α spin density in the ligand:

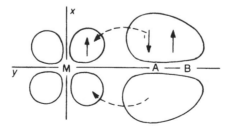

(c) Electron transfer from a full m.o. of the ligand into an empty t_{2g} orbital of the metal, with preference for transfer of α spin so as to increase the number of electrons with the same spin (α) in the metal (Hund's rule):

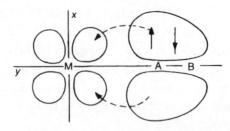

Nuclear spin–spin coupling

The nuclear spin coupling constants in liquids (and gases) are mediated by the electrons. The main contribution is usually the so-called Fermi contact mechanism between the nuclear spins and the electron spins. For such inter-actions the only relevant orbitals are the s orbitals, which are the only ones having non-vanishing density at the corresponding nucleus.

An analytical expression for that contribution can be obtained by using molecular orbital theory at the independent electron level. This was shown by Pople and co-workers (ref. 152) to be directly related (through a constant for each pair of coupled nuclei) to a quantity whose origin lies in the Hückel theory of π systems and named the *mutual polarizability* for the two valence-shell s orbitals centred on the coupled nuclei N and N', $\pi_{ss'}$:

$$\pi_{ss'} = -4 \sum_j \sum_k c_{jsN} c_{ksN} c_{jsN'} c_{ksN'} / (\varepsilon_k - \varepsilon_j). \qquad (12.15)$$

In the above equation, c_{is} is the coefficient of the valence orbital s in the expression of m.o. i and the sums are extended to all the occupied valence m.o.s j and to all empty valence m.o.s k, with energies ε_j and ε_k, respectively.

Replacement of the various $\varepsilon_k - \varepsilon_j$ differences by an effective mean value ΔE converts Eq. (12.15) into Eq. (12.16):

$$\pi_{ss'} = (P_{ss'}^2)^2 / \Delta E \qquad (12.16)$$

where $P_{ss'}$ is the bond order for orbitals s and s', defined in a similar way to Eq. (9.45).

If the coupling between the nuclei in H_2 and HD of necessity involves the m.o.s localized in the bonds H–H and H–D

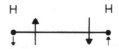

the coupling between nuclei two or more formal bonds away can only be explained by invoking m.o. delocalization. This means that, when considering molecular 'orbitals' χ_i quasi-localized in the formal bonds (see Chapter 8), we cannot ignore the residual interactions

$$\langle \chi_i | \hat{H} | \chi_j \rangle \neq 0 \tag{12.17}$$

in order to explain the coupling through the Fermi contact mechanism. Thus, NMR spectroscopy, like photoelectron spectroscopy, provides the most direct evidence for the non-localizability of m.o.s, both π and σ, a feature that cannot be ignored when properties related to the energy of the electrons are being discussed.

   In valence-bond theory, in terms of which the nuclear spin coupling was first interpreted by the American chemist Martin Karplus (ref. 153), the delocalization of σ electrons is dealt with by including resonant structures such as

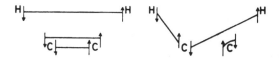

for an HC–CH molecular fragment, in addition to the perfect pairing structure

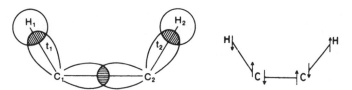

A residual bond order between two hybrid orbitals of C

contributes a positive value to the coupling constant $^3J_{HH}$ between the two vicinal H atoms (three formal bonds away) in the following manner: because of that bonding interaction, the electrons associated with each orbital tend to have opposite spins; this, in turn, favours opposite spins of the s electrons of each H atom (Pauli principle), which, by the Fermi mechanism, favours opposite spins of the H nuclei. Opposite spins of H nuclei mean a positive coupling constant.

The small interaction between the two C hybrid orbitals, as well as the other orbital interactions, can be related to the corresponding overlap integrals and these depend on the geometry of the HC–CH fragment. In particular, the dihedral angle ϕ between the H(1)CC and CCH(2) planes is an important factor. As a consequence, vicinal coupling constants are of extreme value in conformational studies. The relationship (the *Karplus equation*) is of the following type:

$$^3J_{HH} = A + B\cos\phi + C\cos 2\phi. \tag{12.18}$$

Problem 12.14   Using $A = 7\,Hz$, $B = -1\,Hz$ and $C = 5\,Hz$ in Eq. (12.18), determine the ratio between $^3J_{HH}$ for a trans conformation ($\phi = 180°$) and for a gauche conformation ($\phi = 60°$) and compare with the average value of $^3J_{HH}((J_{trans} + 2J_{gauche})/3)$ in ethane: 8.0 Hz.

Problem 12.15   Considering that $^3J_{HH} = 2.6\,Hz$ (for pH between 1.5 and 4.0) and $^3J_{HH} = 3.1\,Hz$ (for pH from 6 to 13) for meso-tartaric acid HO_2C–CH(OH)–CH(OH)–CO_2H, conclude which of the staggered conformations is preferred by the acid and its dinegative ion:

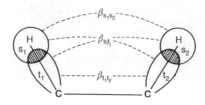

In molecular orbital theory, the long-range residual interactions between otherwise localized and independent m.o.s are described by long-range resonance integrals β:

Just as

$$\beta = \langle t_1|\hat{H}|s_1\rangle \tag{12.19}$$

is a measure of the C–H(1) bonding, the integrals

$$\beta_{t_1 t_2} = \langle t_1 | \hat{H} | t_2 \rangle$$
$$\beta_{t_i s_j} = \langle t_i | \hat{H} | s_j \rangle \qquad (12.20)$$
$$\beta_{s_1 s_2} = \langle s_1 | \hat{H} | s_1 \rangle$$

reflect the residual interactions between the molecular 'orbitals' otherwise localized in C(1)–H(1) and C(2)–H(2). Whereas $\beta_{t_1 t_2}$ contributes a positive value to $^3J_{HH}$, as already shown, the interaction $\beta_{t_i s_j}$ contributes a negative value, that is, favours identical spins of the H nuclei:

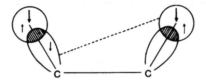

Problem 12.16   What is the sign of the contribution of $\beta_{s_1 s_2}$ to $^3J_{HH}$?

The expression obtained for $^3J_{HH}$ of an HC–CH fragment (ref. 154) is:

$$^3J_{HH} = 392(4\beta_{t_i s_j}^2 - \beta_{s_1 s_2}^2 - 5\beta_{t_1 t_2}^2 + 2\beta_{s_1 s_2}\beta_{t_1 t_2})/\beta^3 + \text{higher order terms}$$
$$(12.21)$$

(with β in eV). Figure 12.10 shows the variation of each term with the dihedral angle and Fig. 12.11 is the resulting Karplus-type curve.

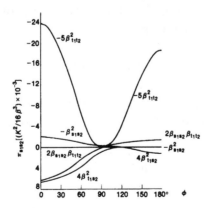

Fig. 12.10 Variation of the terms of Eq. (12.21) with the dihedral angle of an HC–CH fragment (see also ref. 153).

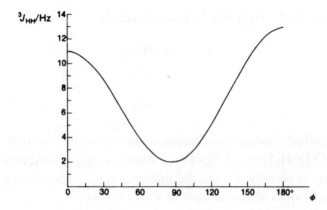

Fig. 12.11 Karplus-type curve of $^3J_{HH}$ for an HC–CH fragment using m.o. theory.

Problem 12.17   The overlap integrals for Slater sp^2 *hybrid orbitals in ethylene are:*

	trans	cis
$S_{t_1 t_2}$	−0.138	0.129
$S_{t_1 s_2}$	−0.031	0.069

By taking $\beta = -10\,SeV$, *determine the ratio* $J_{HH}(cis)/J_{HH}(trans)$ *(ignore the small contribution of* $\beta_{s_1 s_2}$*) and compare with the experimental ratio* 11.5/19.1.

The residual interactions (12.17) between localized molecular 'orbitals' are important even in the case of coupling between directly bonded atoms, such as ^1H and ^{13}C. The $^1J_{CH}$ values for the non-strained hydrocarbons CH_4, C_2H_4, and C_2H_2 are, respectively, 125 Hz, 156 Hz and 248 Hz. These values are almost exactly proportional to the s-character ρ of the hybrid orbitals: sp^3 ($\rho = 1/4$), sp^2 ($\rho = 1/3$) and sp ($\rho = 1/2$):

$$^1J_{CH} \cong 500\,\rho. \tag{12.22}$$

This simple relationship points to an interpretation based on a localized approach involving only the two bonding electrons of the CH bond:

The greater the s-character of the hybrid orbital involved in the bond, the greater the coupling via the Fermi contact mechanism. However, it has been shown (ref. 155) that Eq. (12.22) is the result of a fortuitous cancellation of two neglected effects: (a) to take $\varepsilon_k - \varepsilon_j$ in expression (12.15) as constant; and (b) to ignore residual interactions such as $X = \beta_{st}$ which contribute a negative value to $^1J_{CH}$:

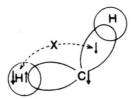

In fact, by choosing the hybrid orbitals so as to maximize orbital overlap and taking $\varepsilon_k - \varepsilon_j$ as constant, the variation with s-character becomes

$$^1J_{CH} \propto \rho^{3/2} \tag{12.23}$$

in agreement with experiment (Fig. 12.12) (ref. 156).

These considerations have been frequently ignored in the literature and in teaching (see, for example, ref. 157), often in association with an incorrect view of orbital hybridization taken as a real phenomenon.

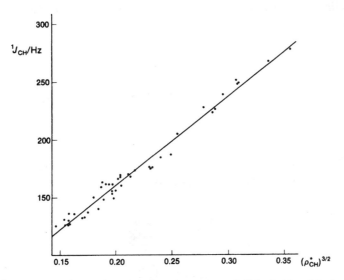

Fig. 12.12 $^1J_{CH}$ for hydrocarbons and s-character of C hybrid orbitals defined by the maximum overlap criterion (see also ref. 156).

Residual interactions between otherwise localized orbitals are also at the root of substituent effects on coupling constants (ref. 158). In the systems

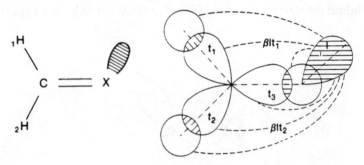

the hyperconjugation of the non-bonding electron pair with CH(1) and CH(2) leads to an increase of $^1J_{CH(1)}$ and a decrease of $^1J_{CH(2)}$, as illustrated by the following examples:

$$^1J_{CH} = 151 \text{ Hz} \qquad ^1J_{CH} = 170 \text{ Hz}$$

These stereospecific lone pair effects are found in many other systems involving various nuclei (ref. 159).

In a similar way, an electronegative atom X, attracting electrons from a CH_2 group,

increases the geminal coupling constant $^2J_{HH}$. The same thing occurs with the hyperconjugation with electron transfer in the opposite direction:

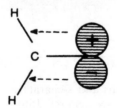

For example, $^2J_{HH}$ = 42.2 Hz in formaldehyde, to be compared with the value 2.3 Hz in ethylene. On the contrary, if the electron donation originates in a non-bonding pair of σ symmetry

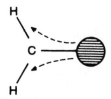

then a negative effect on $^2J_{HH}$ is obtained. The explanation lies in the symmetry of the orbitals involved as acceptors of electrons from the lone pairs. In the case of a π lone pair, electron density is transferred to the asymmetrical anti-bonding molecular 'orbital' χ_4:

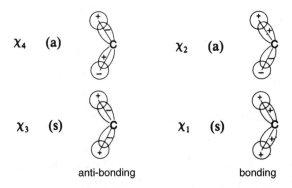

whereas in the case of a σ lone pair it is the symmetrical anti-bonding 'orbital' χ_3 which is involved (ref. 160). These results constitute another neat example of symmetry effects associated with the orbital concept.

Answers to problems

Chapter 2

2.1 $d(\sin x)/dx = \cos x \neq o \sin x$ and $d(\sin x + \cos x)/dx =$
$\cos x - \sin x \neq o (\sin x + \cos x)$. However, $d(\cos x - i \sin x)/dx =$
$-\sin x - i \cos x = -i(\cos x - i \sin x)$, or $d(e^{-ix})/dx = -i e^{-ix}$.

2.2 $d(c_1\psi_1 + c_2\psi_2)/dx = c_1 d\psi_1/dx + c_2 d\psi_2/dx$. However, $\log(c_1\psi_1 + c_2\psi_2) \neq$
$c_1 \log \psi_1 + c_2 \log \psi_2$ and $(c_1\psi_1 + c_2\psi_2)^{1/2} \neq c_1(\psi_1)^{1/2} + c_2(\psi_2)^{1/2}$.

2.3 The scalar product of two perpendicular vectors is zero.

2.4 The integrals $\langle\psi_1|\hat{O}|\psi_2\rangle = \langle\psi_1|o_2\psi_2\rangle = o_2\langle\psi_1|\psi_2\rangle$ and $(\langle\psi_2|\hat{O}|\psi_1\rangle)^* =$
$(\langle\psi_2|o_1\psi_1\rangle)^* = o_1(\langle\psi_2|\psi_1\rangle)^* = o_1\langle\psi_1|\psi_2\rangle$ can only be equal
(expression 2.9), with $o_1 \neq o_2$, if $\langle\psi_1|\psi_2\rangle = 0$.

2.5 $\langle\varphi_i|\hat{A}|\varphi_j\rangle = \langle\varphi_i|a_j\varphi_j\rangle = a_j\langle\varphi_i|\varphi_j\rangle = 0$, because φ_i and φ_j are orthogonal
functions.

2.6 $d[f^2(x)]/dx = 2f(x) d[f(x)]/dx$ which is different from $\{d[f(x)]/dx\}^2$.
The two operators do not commute.

2.7 $[x, d/dx] f(x) = x df(x)/dx - d[x f(x)]/dx = x df(x)/dx - f(x) - x df(x)/dx$
$= -f(x)$; hence, $[x, d/dx] = -1$. $[y, d/dx] f(x) = y df(x)/dx - d[yf(x)]/dx$
$= y df(x)/dx - 0 \times f(x) - y df(x)/dx = 0$; hence, $[y, d/dx] = 0$. The
operators y and d/dx commute.

2.8 $\lambda = 2\pi/k = 2\pi\hbar/(2mE)^{1/2} = h/mv$ and $p = mv = h/\lambda$.

2.9 From Eq. (2.50) we have $d^2\psi/dx^2 = -k^2(Ae^{ikx} + Be^{-ikx}) = -k^2\psi$.
Substitution in Eq. (2.49) leads to $(k^2\hbar^2/2m) \psi = E\psi$, hence Eq. (2.51).

2.10 $\psi = Ae^{ikx} + Be^{-ikx} = A(\cos kx + i \sin kx) + B(\cos kx - i \sin kx) =$
$i(A - B)\sin kx + (A + B)\cos kx$.

2.11 For example, the product $\psi_1\psi_2$ for points between 0 and $a/2$ is symmetrical with the product $\psi_1\psi_2$ for points between $a/2$ and a.

2.12 (a) The energy levels become closer to each other.
 (b) The energy level separations are smaller for higher molecular masses.
 (c) For larger molar volume (or larger molar mass) the higher energy levels become more accessible and, as a consequence, the distribution of molecules among energy levels becomes more disperse: greater molar entropy.

2.13 (a) From Eq. (2.51), with $E - V < 0$ instead of E, k is imaginary and the exponents in Eq. (2.50) become real: $ik = k' = [2m(V - E)]^{1/2}/\hbar$.
 (b) With $A \neq 0$, the first term in $\psi = Ae^{k'x} + Be^{-k'x}$ would go to infinity as $x \to \infty$, whereas the second term vanishes.

2.14 The values $S_z = \pm\hbar/2$ from $S_z = m_s\hbar$ with $m_s = \pm1/2$ are shown as diagonal elements of the matrix $[S_z] = (\hbar/2)[\sigma_z]$.

Chapter 3

3.1 (a) $(-\hbar^2/2m)(\partial^2\psi/\partial x^2 + \partial^2\psi/\partial y^2) = E\psi$.
 (b) $x = r\cos\phi$; $y = r\sin\phi$.
 (c) Using (b), $\partial^2/\partial x^2 + \partial^2/\partial y^2 = (1/r^2)d^2/d\phi^2$ and the wave equation becomes $(-\hbar^2/2mr^2)d^2\psi/d\phi^2 = E\psi$, hence $\hbar^2d^2\psi/d\phi^2 = 2mr^2E\psi = -2IE\psi$ for constant r.
 (d) $J_z = (2IE)^{1/2} = m_\ell\hbar$.

3.2 (a) $J_z = m_\ell\hbar$ and $J = [\ell(\ell+1)]^{1/2}\hbar$.
 (b) $E = J^2/2I = \ell(\ell+1)\hbar^2/2I$.

3.3 For example, using $Y(\theta, \phi) = (3/4\pi)^{1/2}\cos\theta$ in Eq. (3.13), one obtains $-2\cos\theta + \mu\cos\theta = 0$ which requires $\mu = 2$ (that is $\ell = 1$ in Eq. (3.16)).

3.4 Using $\varepsilon_0 = 1/4\pi$ and $Z = 1$, Eq. (3.21) becomes $E = -2\pi^2me^4/n^2\hbar^2$.

3.5 For example, for $n = 4$, there is one s orbital (4s), three p orbitals (4p), five d orbitals (4d) and seven f orbitals (4f), respectively, for $\ell = 0, 1, 2, 3$.

3.6 For example, for $\ell = 1$, the angular parts are $\cos\theta$ (for $m_\ell = 0$) and $\sin\theta\cos\phi$ and $\sin\theta\sin\phi$ (for the pair of values $m_\ell = \pm 1$) which correspond to $\sin\theta\,e^{\pm i\phi} = \sin\theta(\cos\phi \pm i\sin\phi)$ (see also expression 3.41).

3.7 From Table 3.1 we see that the nodal surfaces of the 3s orbital correspond to the roots of the equation $27 - 18\rho + 2\rho^2 = 0$, which are $\rho_1 = 1.9$ and $\rho_2 = 7.1$. For $Z = 1$, with $a_0 = 52.9\,\text{pm}$, this means the existence of two nodal spherical surfaces having $r_1 = 100\,\text{pm}$ and $r_2 = 376\,\text{pm}$.

3.8 Applying Eq. (3.31), one gets $\langle r\rangle_{2s} = 6a_0$ and $\langle r\rangle_{2p} = 5a_0$.

3.9 dP/dr for $P = (1/8)r^2(Z/a_0)^3(2 - Zr/a_0)^2\,e^{-Zr/a_0}$ vanishes for $r = (3 + \sqrt{5})a_0/Z$ (and $d^2P/dr^2 < 0$ for this value of r).

3.10 See Figs. 3.12 and 3.13.

3.11 Using Eq. (3.44) it is found that $\int\psi_{sp}^2\,dv = \int\psi_{sp}'^2\,dv = 1$ because both ψ_{2s} and ψ_{2p} are normalized and $\int\psi_{2s}\psi_{2p}\,dv = 0$. On the other hand, $\int\psi_{sp}\psi_{sp}'\,dv = (1/2)(\int\psi_{2s}^2\,dv - \int\psi_{2p}^2 dv) = (1/2)(1 - 1) = 0$.

3.12 For $\ell = 2$, it is $j = 2 + 1/2 = 5/2$ and $2 - 1/2 = 3/2$.

Chapter 4

4.1 $\int\sigma_1^2\,dv = (\int\varphi_{1sA}^2\,dv + \int\varphi_{1sB}^2\,dv + 2\int\varphi_{1sA}\varphi_{1sB}\,dv)/(2 + 2S) = (1 + 1 + 2S)/(2 + 2S) = 1$.

4.2 $\int\sigma_1'^2\,dv = (\int\varphi_{1sA}^2\,dv + \int\varphi_{1sB}^2\,dv - 2\int\varphi_{1sA}\varphi_{1sB}\,dv)N_1'^2 = 1$, that is $(1 + 1 - 2S)N_1'^2 = 1$, hence $N_1' = 1/[2(1 - S)]^{1/2}$.

4.3 $E_1' = \langle\sigma_1'|\hat{H}|\sigma_1'\rangle = \langle s_A - s_B|\hat{H}|s_A - s_B\rangle/[2(1 - S)] = (\alpha + \alpha - \beta - \beta)/[2(1 - S)] = (\alpha - \beta)/(1 - S)$.

4.4 For the anti-bonding m.o. the energy is $(\alpha - \beta)/(1 - S) = (-4/0.4)\,\text{eV} = -10\,\text{eV}$, which means a stabilization of $(-10 + 4)\,\text{eV} = -6\,\text{eV}$. For the bonding m.o. it is $(\alpha + \beta)/(1 + S) = (-24/1.6)\,\text{eV} = -15\,\text{eV}$, which means a destabilization of $(-15 + 24)\,\text{eV} = +9\,\text{eV}$. Relative to $\alpha = -14\,\text{eV}$, the bonding m.o. lies 1 eV below, in energy, whereas the anti-bonding m.o. lies 4 eV above.

4.5 The larger S the more negative is β and this tends to make the bonding m.o. more stable (and the anti-bonding m.o. more unstable); however, this effect is opposed by an increase of the denominator in $(\alpha + \beta)/(1 + S)$. In addition, if the increase in S results from a smaller R_{AB} value, then not only does the nuclear repulsion increases but also the bonding effect tends to decrease due to a smaller internuclear region where the bonding effect is in part originated. In conclusion, the sentence is controversial.

4.6 The functions (4.18) do not change sign upon inversion through the centre of the molecule (the mid-point in the internuclear line).

4.7 As in Eq. (4.14) and in Problem 4.3, the energies are $(\alpha+\beta)/(1+S)$ and $(\alpha-\beta)/(1-S)$ for the bonding and anti-bonding m.o.s, respectively, with $\alpha = \langle y_A|\hat{H}|y_A\rangle = \langle y_B|\hat{H}|y_B\rangle = \langle z_A|\hat{H}|z_A\rangle = \langle z_B|\hat{H}|z_B\rangle$ and $\beta = \langle y_A|\hat{H}|y_B\rangle = \langle z_A|\hat{H}|z_B\rangle$.

Chapter 5

5.1
$$\begin{vmatrix} 1s(1) & 1s(2) & 1s(3) \\ \overline{1s}(1) & \overline{1s}(2) & \overline{1s}(3) \\ 2s(1) & 2s(2) & 2s(3) \end{vmatrix} \quad \text{and} \quad \begin{vmatrix} 1s(1) & 1s(2) & 1s(3) \\ \overline{1s}(1) & \overline{1s}(2) & \overline{1s}(3) \\ \overline{2s}(1) & \overline{2s}(2) & \overline{2s}(3) \end{vmatrix}$$

5.2 $ad - bc = ad + \lambda ac - bc - \lambda ac$ for any value of λ.

5.3 The maximum of the radial probability distribution occurs for a higher r value in the SCF case, due to the inclusion of the (averaged) electronic repulsion.

5.4 For 3d orbitals which are more penetrating than 3p and 3s.

5.5 No, because the orbitals $1s, 2s, 2p, \ldots$ change with Z (they become more compact around the nucleus as Z increases).

5.6 (a) Na: $1s^2, 2s^2, 2p^6, 3s^1$ with $\sigma_{3s} = 8 \times 0.85 + 2 \times 1.00 = 8.80$ and $\sigma_{2p} = 7 \times 0.35 + 2 \times 0.85 = 4.15$.
  (b) $Z^* = Z - \sigma$ increases by $(1.00 - 0.35) = 0.65$ for each unit increment of Z, from $Z = 3$ $(Z^* = 1.3)$ till $Z = 10$ $(Z^* = 5.85)$. For $Z = 11$, $Z^* = 11 - 8.8 = 2.2$.

5.7 O: $1s^2, 2s^2, 2p^4$ and $Z_1^* = 8.00 - (5 \times 0.35 + 2 \times 0.85) = 4.55$. O^+: $1s^2$, $2s^2, 3p^3$ and $Z_2^* = 8.00 - (4 \times 0.35 + 2 \times 0.85) = 4.90$. From Eq. (5.37) one obtains $I_2/I_1 = (Z_2^*/Z_1^*)^2 = 1.16$, a value which is significantly smaller than the experimental one: 2.6.

5.8 First figure relative to the K atom. There is a significant jump of $\log I$ when going from the first ionization (removal of the 4s electron from K) to the second ionization (removal of a 3p electron from K^+), followed by a smooth variation up to the ninth ionization (successive removal of electrons of $n=3$), until a new jump is recorded for the 10th ionization (removal of the first electron with $n=2$); and similarly for the 18th and 19th ionizations (removal of 1s electrons). Second figure relative to atoms of increasing Z: $\sqrt{I}$ changes regularly within each block corresponding to each value of the principal quantum number (see answer above).

5.9 For $1s^2$, $2s^2$, $2p^6$, $3s^2$, $3p^6$, $3d^p$, $4s^q$, we have $\sigma_{4s}=(q-1)\times 0.35+(8+p)\times 0.85+10\times 1.00=0.35q+0.85p+16.45$, a value that increases when q decreases by 1 so that p increases by 1. Similarly, σ_{3d} increases by 0.35 for each additional d electron coming from the 4s orbital (note that this orbital does not contribute to σ_{3d} because it corresponds to a higher n value).

5.10 The difference is $E_{1,1/2,3/2} - E_{1,1/2,1/2}=(1/2)\,hcA\,(15/4-3/4)=$ $(3/2)\times 6.63\times 10^{-34}\,\text{J s}\times 3.00\times 10^{10}\,\text{cm s}^{-1}\times 38.5\,\text{cm}^{-1}=$ $1.15\times 10^{-31}\,\text{J}$.

5.11 The level $3P_{3/2}$ lies above $3P_{1/2}$ by an energy value corresponding to $17\,\text{cm}^{-1}$, that is, $hc\Delta\bar{\nu}=6.63\times 10^{-34}\,\text{J s}\times 3.00\times 10^{10}\,\text{cm s}^{-1}\times$ $17\,\text{cm}^{-1}=3.4\times 10^{-32}\,\text{J}$.

5.12 For $2p^1$, $3p^1$, we have $\ell_1=1$ and $\ell_2=2$, hence L can be 2, 1 or 0. Meanwhile, S can be 1 or 0. Thus, for $S=0$, the levels are: $L=2$, $J=2$, level $^1D_2$; $L=1$, $J=1$, level $^1P_1$; $L=0$, $J=0$, level $^1S_0$. For $S=1$, the levels are: $L=2$, $J=3$, level $^3D_3$; $L=2$, $J=2$, level $^3D_2$; $L=2$, $J=1$, level $^3D_1$; $L=1$, $J=2$, level $^3P_2$; $L=1$, $J=1$, level $^3P_1$; $L=1$, $J=0$, level $^3P_0$; $L=0$, $J=1$, level $^3S_1$.

5.13 For $1s^2$, $2s^2$, $2p^3$ the highest total spin quantum number is $S=3/2$, which corresponds to a spin multiplicity $2S+1=4$, with $\Sigma m_\ell=0$: ground state $^4S_{3/2}$ (application of Hund's first rule). The increase of the ionization energy on going from C to N is mainly due to an increase of the effective nuclear charge. The decrease on going from N to O is mainly due to a reduction of the number of unpaired spins.

5.14 (a) Each of the two determinants in Eq. (5.52) assumes opposite spins.

(b) $\quad (1/2)\begin{vmatrix} 1s(1) & 1s(2) \\ \overline{2s}(1) & \overline{2s}(2) \end{vmatrix} - (1/2)\begin{vmatrix} \overline{1s}(1) & \overline{1s}(2) \\ 2s(1) & 2s(2) \end{vmatrix}$

5.15 The stabilization of the 6s orbital leads to a weaker 6s–6s interaction between neighbouring atoms than would be expected for metals in general. The 6p–6p interactions are also weak. The van der Waals forces are weak interactions. Hence, gaseous mercury is monoatomic and the solidification temperature is abnormally low.

Chapter 6

6.1 For example, the probability density $\sigma_1'^2$ is given by the same expression $(\varphi_{1sA}^2 + \varphi_{1sB}^2 - 2\varphi_{1sA}\varphi_{1sB})/(2 - 2S)$.

6.2 $(1\sigma_g)^2, (1\sigma_u')^2, (1\pi_u)^4, (2\sigma_g)^1$, one less bonding electron than in N_2.

6.3 $(1\sigma_g)^2, (1\sigma_u')^2, (2\sigma_g)^2, (1\pi_u)^4, (1\pi_g')^3$ for O_2^- and $(1\sigma_g)^2, (1\sigma_u')^2, (2\sigma_g)^2,$ $(1\pi_u)^4, (1\pi_g')^4$ for O_2^{2-}.

6.4 $\int \sigma_1 \sigma_1' \, dv = 0.33 \times 0.94 \times 1 - 0.94 \times 0.33 \times 1 - 0.33 \times 0.33 \times 0 + 0.94 \times$ $0.94 \times 0 = 0$ (if the overlap integral between φ_{1sH} and φ_{2pxF} is ignored).

6.5 It would be $\int \sigma_1^2 \, dv = 0.323 \times 0.323 \times 1 + 0.231 \times 0.231 \times 1 + 0.685 \times$ $0.685 \times 1 = 0.626 \neq 1$.

6.6 Four bands corresponding to the removal of one electron from the σ_1, σ_2, π and σ_3 molecular orbitals of Fig. 6.14.

6.7 A diagram similar to Fig. 6.14: $(\sigma_1)^2, (\sigma_2)^2, (\pi)^4, (\sigma_3)^2$.

Chapter 7

7.1 $S(r = 2)/S(r = 3) = 1.7$

7.2 $\int \sigma_2 \sigma_3 \, dv = N_2 N_3 (\int \varphi_{1sA}^2 \, dv - \lambda \int \varphi_{1sA} \varphi_{1sB} \, dv + \int \varphi_{1sA} \varphi_{1sC} \, dv -$ $\int \varphi_{1sA} \varphi_{1sC} \, dv + \lambda \int \varphi_{1sB} \varphi_{1sC} \, dv - \int \varphi_{1sC}^2 \, dv) = N_2 N_3 (1 - \lambda S + S' -$ $S' + \lambda S - 1) = 0$.

7.3 $\int \sigma_1 \sigma_3 \, dv = N_1 N_3 (\int \varphi_{1sA}^2 \, dv - 2 \int \varphi_{1sA} \varphi_{1sB} \, dv + \int \varphi_{1sA} \varphi_{1sC} \, dv +$ $\int \varphi_{1sA} \varphi_{1sB} \, dv - 2 \int \varphi_{1sB}^2 \, dv + \int \varphi_{1sB} \varphi_{1sC} \, dv + \int \varphi_{1sA} \varphi_{1sC} \, dv -$ $2 \int \varphi_{1sB} \varphi_{1sC} \, dv + \int \varphi_{1sC}^2 \, dv) = N_1 N_3 (1 - 2S + S + S - 2 + S +$ $S - 2S + 1) = 0$.

7.4 Since $N_2 = 1/[2(1 - S)]^{1/2}$ and $N_3 = 1/[6(1 - S)]^{1/2}$, then $N_2^2 = 3N_3^2$.

7.5 $\sigma = \sigma_1^2 + \sigma_2^2 = 2(s_1^2 + s_2^2) + a^2 s^2 + b^2 x^2 + 2as(s_1 + s_2)$ is symmetric with respect to the Be nucleus. It is noted that $s_1 x = -s_2 x$.

7.6 $\int s_1 x\, dv = -\int s_2 x\, dv$, hence $\int \phi_s x\, dv = 0$ and $\int \phi_a x\, dv = \int s_1 x\, dv -$
$\int s_2 x\, dv \neq 0$.

7.7 σ_2 for BH_3 and NH_3 is similar to σ_2 for H_2O (see page 145 and Figs. 7.10 and 7.11), but in H_2O (and BeH_2) the centres of the 1s a.o.s of H and the axis of 2p of the central atom lie in the same plane (line).

Chapter 8

8.1 Four local concentrations of valence shell charge, arranged approximately in a tetrahedral way: three equivalent bonded concentrations and a larger non-bonded concentration.

8.2 Eighteen valence m.o.s constructed from $3 \times 4 + 6 = 18$ valence a.o.s: nine bonding m.o.s all filled with the 18 valence electrons; $6 \times 2 = 12$ valence electrons are responsible for the six C–H bonds, there being $18 - 12 = 6$ electrons responsible for two C–C bonds: two electrons for the long C–C bond (a single bond) and four electrons for the short C=C bond (a double bond).

8.3 (a) sp. (b) sp^2. (c) sp^2. (d) sp. (e) sp^2. (f) sp^3.

8.4 (a)

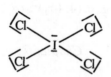

with two non-bonding electron pairs in I which find more space for an axial orientation, in agreement with an equatorial arrangement of the four chlorine atoms.

(b)

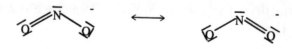

with a bent geometry in agreement with three (effective) electron pairs (domains) around the central atom.

(c) Six electron pairs around the central atom: octahedral geometry.

(d)

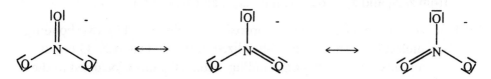

with a trigonal planar geometry in agreement with three (effective) electron pairs (domains) around the N atom.

8.5 A larger size of the double bond domain of ethylene allows the prediction of a HCH angle smaller than 120°, in agreement with experiment. The bonding pair domains in the valence shell of C are smaller in CH_3Cl than in CH_4, as a result of the greater electronegativity of Cl compared to H; an increase of the HCH angle can be predicted, in agreement with experiment.

8.6 It is 3 for any of the functions.

8.7 The ratio of the squared coefficients for p and s is $0.89^2/0.45^2 = 4$.

8.8 Three equivalent molecular 'orbitals' almost localized in the three N–H bonds and an approximately non-bonding molecular 'orbital' almost localized in the N atom. Since the bond angles (107°) are very close to those for the axes of sp^3 atomic orbitals, the non-bonding 'orbital' can be considered approximately as an sp^3 hybrid orbital of N. Each one of the bonding molecular 'orbitals' is approximately the combination of a 1s orbital of H with an sp^3 orbital of N.

8.9 The localized functions which describe the C–H and C–C bonds are not canonical m.o.s; no energy can be rigorously associated with them.

Chapter 9

9.1 (a) $\langle \chi_1|\hat{H}|\chi_2\rangle = \langle sp_1|\hat{H}|sp_2\rangle + \langle x_{O(1)}|\hat{H}|x_{O(2)}\rangle + \langle sp_1|\hat{H}|x_{O(2)}\rangle + \langle x_{O(1)}|\hat{H}|sp_2\rangle$.

(b) $\langle sp_1|\hat{H}|sp_2\rangle = \langle s|\hat{H}|s\rangle - \langle p|\hat{H}|p\rangle - \langle s|\hat{H}|p\rangle + \langle p|\hat{H}|s\rangle = \langle s|\hat{H}|s\rangle - \langle p|\hat{H}|p\rangle \neq 0$, because the energies of the 2s orbital and 2p orbitals are different.

9.2 The sum of the overlap integrals between z_C and $z_{O(1)}$ and $z_{O(3)}$ is cancelled by $-2\langle z_C|z_{O(2)}\rangle$.

9.3 For example, the combination $2\pi_1 + b(\phi_2 + \phi_3)$ has no contributions from $z_{O(2)}$ and $z_{O(3)}$ but includes $z_{O(1)}$ and $z_{O(3)}$, besides z_C.

9.4 (a) A σ component which is approximately described by two bonding molecular 'orbitals' quasi-localized in the O–O and N–O bonds and five essentially non-bonding 'orbitals' almost localized in the various atoms (one in the central atom and two × two in the terminal atoms); in view of a value close to 120° for the bond angle, sp^2 hybrid orbitals can be chosen. A delocalized (and non-localizable) π component constructed from three 2p orbitals having parallel axes, one from each atom: one bonding π m.o., one anti-bonding and one non-bonding.

(b) Three effective electron pairs (domains) around the central atom.

9.5 The $H_2C{=}O$ molecule can be imagined to be obtained from $H_2C{=}CH_2$ by associating the nuclei of the two H atoms bonded to one of the C atoms with the nucleus of this atom; the atomic number of the latter becomes 8 (oxygen isotope). The bonding molecular 'orbitals' localized in the corresponding C–H bonds become non-bonding, localized in O: $H_2C{=}\overline{O}$. The remaining molecular 'orbitals' (σ and π) do not change significantly, except for the polarity of the CO bond.

9.6 See answer to Problem 4.7.

9.7 The HC≡CH molecule can be imagined to be obtained from N≡N by letting one proton out of each N nucleus, in opposite directions along the internuclear line, up to a certain distance (the C–H bond length); the atomic number of the initial nuclei becomes 6 (carbon isotope) and the removed protons become nuclei of H atoms. The two non-bonding molecular 'orbitals' localized in the N atoms are converted into bonding C–H molecular 'orbitals'. The form of the remaining molecular 'orbitals' (σ and π) does not change significantly.

9.8 $\int (c_1 z_1 + c_2 z_2 + c_3 z_3 + c_4 z_4)^2 \, dv = 1 = c_1^2 \langle z_1 | z_1 \rangle + c_2^2 \langle z_2 | z_2 \rangle + c_3^2 \langle z_3 | z_3 \rangle + c_4^2 \langle z_4 | z_4 \rangle + \Sigma c_i c_j \langle z_i | z_j \rangle = c_1^2 \times 1 + c_2^2 \times 1 + c_3^2 \times 1 + c_4^2 \times 1 + \Sigma c_i c_j \times 0 = c_1^2 + c_2^2 + c_3^2 + c_4^2.$

9.9 Use $x = -0.62$ in the secular equations (9.29) and solve as was done for π_1.

9.10 The coefficients in Eq. (9.32) satisfy the condition (9.31). On the other hand $\langle \pi_1 | \pi_2 \rangle = 0.37 \times 0.60 + 0.60 \times 0.37 - 0.60 \times 0.37 - 0.37 \times 0.60 = 0.$

9.11 The energy levels (translational energy) given by Eq. (2.60) decrease when the size of the box (a) increases.

9.12 For ethylene, we have $\alpha + \beta$ and $\alpha - \beta$. In the case of butadiene, we have $\alpha + 1.62\beta$ and $\alpha - 1.62\beta$ for π_1 and π_4, respectively, and $\alpha + 0.62\beta$ and $\alpha - 0.62\beta$ for π_2 and π_3, respectively: two pairs of m.o.s where the absolute values of the coefficients are the same and change sign for alternate a.o.s (see expressions (9.32) and (9.33)).

9.13 One bonding m.o. of energy $\alpha + 2\beta$ and two non-bonding m.o.s of energy α (besides an anti-bonding m.o. of energy $\alpha - 2\beta$). A sum of energies for the four π electrons which is equal to the value for two separate C=C bonds: $4\alpha + 4\beta$. Actually, there are alternating single and double bonds in cyclobutadiene.

9.14 By summing the squared coefficients of each z_i in π_1, π_2 and π_3 after multiplication by 2, one obtains 1 for each value of i.

9.15 With $n = 2$ and $c_1 = c_2 = (1/2)^{1/2}$, Eq. (9.45) gives $p_{12} = 1$.

Chapter 10

10.1 (a) The fact that the F–F bond is very weak contributes to negative values of the enthalpy changes of reactions involving F_2 as a reactant.
   (b) The fact that the N–N and N–X bonds are particularly weak contributes to the exothermicity of the reactions of explosives with oxygen.

10.2 The reaction $2O_3 \rightarrow 3O_2$ implies the replacement of two × two oxygen–oxygen bonds having a bond order between 1 and 2 by three × one double bonds. The fact that the O=O bond energy ($494\,\mathrm{kJ\,mol^{-1}}$) is much greater than twice the O–O bond energy ($142\,\mathrm{kJ\,mol^{-1}}$) makes the reaction exothermic.

10.3 (a) The S=S bond is weak, hence structures like S_8 (eight S–S bonds per molecule) are particularly stable relative to four S_2 molecules (a total of four double bonds).
   (b) The S=O bond is much weaker than C=O. Giant covalent structures involving single bonds and having the empirical formula SiO_2 are stable relative to O=Si=O molecules.

10.4 The atomization energy of P_4 is $6 \times 220\,kJ\,mol^{-1} = 1320\,kJ\,mol^{-1}$, a value much greater than twice the bond energy of $P{\equiv}P$ ($481\,kJ\,mol^{-1}$). However, for N_4 the corresponding values would be $6 \times 167\,kJ\,mol^{-1} = 1002\,kJ\,mol^{-1}$ to be compared to $2 \times 942\,kJ\,mol^{-1} = 1884\,kJ\,mol^{-1}$. Phosphorus exists as tetratomic molecules (with nuclei defining the vertices of a tetrahedron) whereas nitrogen exists as diatomic molecules.

10.5 $|N{\equiv}N|, \overline{O} = \overline{O}, |\overline{F}-\overline{F}|, H-\overline{F}|$, and $|C{\equiv}O|$.

10.6 The difference of electronegativity for F and N is greater than for Cl and O. Accordingly, F_2N occurs only as FNF with the two negative electric poles (F) at the ends of the molecule.

10.7 (a) Because of the distribution of the ionic charge in $N_3{}^-$, the hypothetical addition of a proton to a terminal N nucleus of $N_3{}^-$ is expected to lead to greater stabilization than if it was added to the central nucleus.

(b) In order to 'obtain' NON from CO_2 one proton from each O nucleus would have to be moved into the central nucleus, whereas only one such transfer is necessary to 'obtain' NNO; the latter is less unfavourable (consider the negative electric poles located at the O atoms in CO_2).

10.8 1,3-Diazine and 1,3,5-triazine, with (electronegative) N atoms instead of CH units at the positions of negative charge in the ring of pyridine.

10.9 (a) 6.   (b) 1.   (c) $\overline{N}-\overline{N}$.

Chapter 11

11.1 From Eq. (11.1) and using $R = 350\,pm$, we get $E_5 - E_4 = 10h^2/8\pi^2mR^2 = 4.99 \times 10^{-19}\,J = hc/\lambda$, hence $\lambda = 398\,nm$.

11.2 For example, $\langle n_5 - n_6 | 2n_1 - n_2 - n_3 + 2n_4 - n_5 - n_6 \rangle = 2\langle n_5 | n_1 \rangle - \langle n_5 | n_2 \rangle - \langle n_5 | n_3 \rangle + 2\langle n_5 | n_4 \rangle - \langle n_5 | n_5 \rangle - \langle n_5 | n_6 \rangle - 2\langle n_6 | n_1 \rangle + \langle n_6 | n_2 \rangle + \langle n_6 | n_3 \rangle - 2\langle n_6 | n_4 \rangle + \langle n_6 | n_5 \rangle + \langle n_6 | n_6 \rangle = 2S - S - S + 2S - 1 - S - 2S + S + S - 2S + S + 1 = 0.$

11.3 (a) They do not change sign upon inversion through the centre of the atom.

(b) The p orbitals (t_{1u}) change sign upon inversion through the centre of the atom, whereas the d_{xy}, d_{xz}, and d_{yz} orbitals (t_{2g}) do not.

11.4 Fe(III) having the sub-configuration $3d^5$: $(t_{2g})^5$ with one unpaired spin (low spin) or $(t_{2g})^3$, $(e'_g)^2$ with five unpaired electrons (high spin). Co(II) having the sub-configuration $3d^7$: $(t_{2g})^6$, $(e'_g)^1$ with one unpaired spin (low spin) or $(t_{2g})^5$, $(e'_g)^2$ with three unpaired spins (high spin).

11.5 Since $[Ni(H_2O)_6]^{2+}$ is a weak field complex, the low-lying e'_g m.o.s occupied by two unpaired electrons contribute to stabilization (Hund's rule).

11.6 Identical to Eq. (7.4) (bonding), Eq. (7.5) (non-bonding) and Eq. (7.6) (anti-bonding).

11.7 The same interpretation as for Eq. (9.24) (butadiene), except that the Hückel Coulomb and resonance integrals are now defined in terms of two s orbitals instead of p orbitals of π symmetry.

11.8 $\beta' = (1/4)\langle s + x + y + z|\hat{H}|s - x + y - z\rangle = (1/4)[\langle s|\hat{H}|s\rangle - \langle s|\hat{H}|x\rangle + \langle s|\hat{H}|y\rangle - \langle s|\hat{H}|z\rangle + \langle x|\hat{H}|s\rangle - \langle x|\hat{H}|x\rangle + \langle x|\hat{H}|y\rangle - \langle x|\hat{H}|z\rangle + \langle y|\hat{H}|s\rangle - \langle y|\hat{H}|x\rangle + \langle y|\hat{H}|y\rangle - \langle y|\hat{H}|z\rangle + \langle z|\hat{H}|s\rangle - \langle z|\hat{H}|x\rangle + \langle z|\hat{H}|y\rangle - \langle z|\hat{H}|z\rangle] = (1/4)(\alpha_s - \alpha_p)$.

Chapter 12

12.1 The nitro group removes electron density from the ring.

12.2 The interaction of the terminal parts of the HOMO having anti-bonding character contributes to make the *cis* form unstable relative to *trans*.

12.3 (a) The overlap integral between the HOMO of one molecule and the LUMO of the other is zero.
  (b) There is a non-zero overlap between the HOMO of the excited molecule (π') and the LUMO of the other (π).

12.4 The planar structure provides delocalization of the π system, hence $\pi \rightarrow \pi'$ transitions of lower energy which can be accomplished by absorption of visible light.

12.5 According to expression (2.62) the separation between successive energy levels decreases when the size of the box increases.

12.6 Steric interactions between the methyl groups in 2,2'-dimethylbiphenyl lead to a larger average angle between the planes of the two benzene rings, hence π electron delocalization is reduced relative to the 3,3'- and 4,4'-isomers.

12.7   The interaction of the non-bonding 'orbitals' of pyridazine, n_1 and n_2, can be interpreted in terms of two m.o.s, $n_1 + n_2$ and $n_1 - n_2$, both filled, one of lower energy ($n_1 + n_2$) and the other of higher energy ($n_1 - n_2$). $n \rightarrow \pi'$ transitions from the latter require less energy.

12.8   A shift to lower frequencies because of delocalization of the π system in benzaldehyde.

12.9 $[Co(NH_3)_6]^{3+}$.

12.10 The transition dipole moment in any direction, x for example ($d_{1s,2s} = \langle 1s|-exi|2s \rangle$) is zero, because the integrand has symmetric values for $x > 0$ and $x < 0$.

12.11 The probability of $t_{2g} \rightarrow e'_g$ transitions is small as a result of transitions between pure d orbitals being prohibited (the selection rule for allowed transitions is $\Delta \ell = \pm 1$).

12.12 A π electron density larger than 1 at the *para* position.

12.13 By application of Eq. (12.14), $\Delta \rho_\pi = -0.036$. Therefore, $\rho_\pi = 1 - 0.036 = 0.964$.

12.14 For $\phi = 180°$, $J_{trans} = 13\,Hz$. For $\phi = 60°$, $J_{gauche} = 4\,Hz$. Ratio $J_{trans}/J_{gauche} = 3.3$. The calculated average value for $^3J_{HH}$ is $(13 + 2 \times 4)/3\,Hz = 7\,Hz$, in good agreement with experiment.

12.15 The conformation where the H atoms are *gauche*, because the coupling constant is typical of a *gauche* arrangement.

12.16 A positive contribution: it favours an antiparallel alignment of the nuclear spins, as in H_2 (see page 282).

12.17 By application of Eq. (12.21), we have $(J_{cis}/J_{trans}) = (4 \times 0.069^2 - 5 \times 0.129^2)/[4 \times (-0.031)^2 - 5 \times (-0.138)^2] = 0.70$, to be compared with the experimental data: $11.5/19.1 = 0.60$.

References

1. K.J. LAIDLER, The story of quantum chemistry, part 1. *Chem 13 News*, no. 191 (1990), and *The World of Physical Chemistry*, Oxford University Press, UK (1993).
2. B.L. HAENDLER, Presenting the Bohr atom. *J. Chem. Educ.*, **59**, 372 (1982).
3. M.A. MORRISON, *Understanding Quantum Physics*, Prentice Hall, USA (1990).
4. P.M. DIRAC, *Principles of Quantum Mechanics*, fourth edition, Clarendon Press, UK (1958).
5. P.G. MERLI, G.F. MISSORI and G. POZZI, On the statistical aspect of electron interference phenomena. *Am. J. Phys.*, **44**, 306 (1976).
6. T.F. JORDAN, Disappearance and reappearance of macroscopic quantum interference. *Phys. Rev. A*, **48**, 2449 (1993).
7. R.S. TREPTOW, Precision and accuracy in measurements. *J. Chem. Educ.*, **75**, 992 (1998).
8. F. CASTAÑO, L. LAIN, M.N. SANCHEZ RAYO and A. TORRE, Does quantum mechanics apply to one or many particles? *J. Chem. Educ.*, **60**, 377 (1983).
9. G. HENDERSON, 2-D wave packets and the Heisenberg uncertainty principle. *J. Chem. Educ.*, **70**, 972 (1993).
10. D.F. STYER, Common misconceptions regarding quantum mechanics. *Am. J. Phys.*, **64**, 31 (1996).
11. O.G. LUDWIG, On a relation between the Heisenberg and de Broglie principles. *J. Chem. Educ.*, **70**, 28 (1993).
12. P.W. ATKINS, *Molecular Quantum Mechanics*, Oxford University Press, UK (1983) and *Physical Chemistry*, fifth edition, Oxford University Press, UK (1994).
13. J.P. LOWE, *Quantum Chemistry*, second edition, Academic Press, USA (1993).
14. J.J.C. TEIXEIRA DIAS, How to teach the postulates of quantum mechanics without enigma. *J. Chem. Educ.*, **60**, 963 (1983).
15. T.L. CHOW, A note on the Schrödinger representation of the momentum and energy operators. *J. Chem. Educ.*, **69**, 537 (1992).
16. P.W. ATKINS, *Quanta, A Handbook of Concepts*, second edition, Oxford University Press, UK (1994).
17. D. BRESSANINI and A. PONTI, Angular momentum and the two-dimensional free particle. *J. Chem. Educ.*, **75**, 916 (1998).

18. Y.Q. LIANG, H. ZHANG and Y.X. DARDENNE, Momentum distributions for a particle in a box. *J. Chem. Educ.*, **72**, 148 (1995).
19. F. RIOUX, Numerical methods for finding momentum space distributions. *J. Chem. Educ.*, **74**, 605 (1997).
20. F. RIOUX, Electron-momentum spectroscopy and the measurement of orbitals: interesting results for chemists from the *American Journal of Physics*. *J. Chem. Educ.*, **76**, 156 (1999) and references therein.
21. P.G. NELSON, About electrons crossing nodes. *J. Chem. Educ.*, **70**, 345 (1993).
22. P.L. LANG and M.H. TOWNS, Visualization of wave functions using Mathematica. *J. Chem. Educ.*, **75**, 506 (1998).
23. *Mathematica 2.0*, Wolfram Research, Champaign, IL 61820-7237, USA.
24. J.C.A. BOEYENS, Understanding electron spin. *J. Chem. Educ.*, **72**, 412 (1995).
25. P.V. ALBERTO, C. FIOLHAIS and V.M.S. GIL, Relativistic particle in a box. *Eur. J. Physics*, **17**, 19 (1996).
26. H. EYRING, J. WALTER and G.E. KIMBALL, *Quantum Chemistry*, John Wiley, USA, (1967).
27. F. RIOUX, Enriching quantum chemistry with Mathcad, part II. *J. Chem. Educ.*, **72**, 327 (1995).
28. F. RIOUX, Enriching quantum chemistry with Mathcad (for Macintosh). *J. Chem. Educ.*, **74**, 1016 (1997).
29. M. VOS and I. McCARTHY, Measuring orbitals and bonding in atoms, molecules and solids. *Am. J. Phys.*, **65**, 544 (1997).
30. R.D. ALLENDOERFER, Teaching the shapes of the hydrogenlike and hybrid atomic orbitals. *J. Chem. Educ.*, **67**, 37 (1990).
31. R. BART, Where the electrons are? *J. Chem. Educ.*, **72**, 401 (1995).
32. B. RAMACHANDRAN and P.C. KONG, Three-dimensional graphical visualization of one-electron atomic orbitals. *J. Chem. Educ.*, **72**, 406 (1995).
33. B. RAMACHANDRAN, Examining the shapes of atomic orbitals using Mathcad. *J. Chem. Educ.*, **72**, 1082 (1995).
34. R.E. POWELL, Relativistic quantum chemistry: the electrons and the nodes. *J. Chem. Educ.*, **45**, 558 (1968).
35. L. PISANI, J.-M. ANDRÉ and E. CLEMENTI, Study of relativistic effects in atoms and molecules by the kinetically balanced LCAO approach. *J. Chem. Educ.*, **70**, 894 (1993).
36. P.V. ALBERTO, M. FIOLHAIS and M. OLIVEIRA, On the relativistic $L–S$ coupling. *Eur. J. Phys.*, **19**, 553 (1998).
37. J.R. REIMERS, G.B. BACSKAY and S. NORDHOLM, The basis of covalent bonding (J. Chem. Educ. Software). *J. Chem. Educ.*, **74**, 1503 (1997).
38. M.J. FEINBERG, K. RUEDENBERG and E.L. MEHLER, in *Advances in Quantum Chemistry*, Vol. 5, edited by P.O. Lowdin, Academic Press, New York (1970).
39. G.B. BACSKAY, J.R. REIMERS and S. NORDHOLM, The mechanism of covalent bonding. *J. Chem. Educ.*, **74**, 1494 (1997).
40. J.K. BURDETT, *Chemical Bonds: A Dialog*, John Wiley, UK (1997).
41. J.F. OGILVIE, There are no such things as orbitals. *J. Chem. Educ.*, **67**, 280 (1990).
42. J. SIMONS, There are no such things as orbitals – act two. *J. Chem. Educ.*, **68**, 131 (1991).

43. F. RIOUX, An interactive SCF calculation. *J. Chem. Educ.*, **71**, 781 (1994).
44. L.S. BARTELL and L.O. BROCKWAY, The investigation of electron distributions in atoms by electron diffraction. *Phys. Rev.*, **90**, 833 (1953).
45. M.K. HARBOLA and V. SAHNI, Theories of electronic structure in the Pauli-correlated approximation. *J. Chem. Educ.*, **70**, 920 (1993).
46. C. ANGELI, Physical interpretation of Koopman's theorem: a criticism of the current didactic presentation. *J. Chem. Educ.*, **75**, 1494 (1998).
47. J. SLATER, *Quantum Theory of Atomic Structure*, vol. 1, McGraw Hill, USA (1960).
48. N.E. ROBERTSON, S.O. BOGGIONI and F.E. HENRIQUEZ, A simple mathematical approach to experimental ionization energies of atoms. *J. Chem. Educ.*, **71**, 101 (1994).
49. N.C. PYPER and M. BERRY, Ionization energies revisited. *Educ. in Chem.*, 135 (1990).
50. P. MIRONE, How to get more from ionization energies in the teaching of atomic structure. *J. Chem. Educ.*, **68**, 132 (1991).
51. M.P. MELROSE and E.R. SCERRI, Why is the 4s orbital occupied before the 3d? *J. Chem. Educ.*, **73**, 498 (1996).
52. J.L. BILLS, Experimental 4s and 3d energies in atomic ground states. *J. Chem. Educ.*, **75**, 589 (1998).
53. P.G. NELSON, Relative energies of 3d and 4s orbitals. *Educ. in Chem.*, **29**, 84 (1992).
54. L.G. VANQUICKENBORNE, K. PIERLOT and D. DEVOGHEL, Transition metals and the aufbau principle. *J. Chem. Educ.*, **71**, 469 (1994).
55. M.L. CAMPBELL, The correct interpretation of Hund's rule as applied to 'uncoupled states' orbital diagrams. *J. Chem. Educ.*, **68**, 134 (1991).
56. N. SHENKUAN, The physical basis of Hund's rule. *J. Chem. Educ.*, **69**, 800 (1992).
57. I. NOVAK, Electronic states and configurations: visualizing the difference. *J. Chem. Educ.*, **76**, 135 (1999).
58. G. DOGGETT and B. SUTCLIFFE, A modern approach to *L–S* coupling in the theory of atomic spectra. *J. Chem. Educ.*, **75**, 110 (1999).
59. C.W. HAIGH, The theory of atomic spectroscopy: jj coupling, intermediate coupling and configuration interaction. *J. Chem. Educ.*, **72**, 206 (1995).
60. R.S. TREPTOW, The Periodic Table of atoms. *J. Chem. Educ.*, **71**, 1007 (1994).
61. R.G. PARR and W. YANG, Density-functional theory of the electronic structure of molecules. *Annu. Rev. Phys. Chem.*, **46**, 701 (1995).
62. P. PYYKKO, Relativistic effects in structural chemistry. *Chem. Rev.*, **88**, 563 (1988).
63. A.H. GUERRERO, H.J. FASOLI and L.L. COSTA, Why gold and copper are colored but silver is not. *J. Chem. Educ.*, **76**, 200 (1999).
64. L.J. NORBY, Why is mercury liquid? *J. Chem. Educ.*, **68**, 110 (1991).
65. C.J. WILLIS, Describing electron distribution in the hydrogen molecule. *J. Chem. Educ.*, **68**, 743 (1991).
66. A.D. BAKER and D. BETTERIDGE, *Photoelectron Spectroscopy*, in International Series in Analytical Chemistry, vol. 53, Pergamon Press, UK (1972).
67. A.C. WAHL, Molecular orbital densities. *Science*, **151**, 961 (1966).
68. A. HAIM, The relative energies of molecular orbitals for second-row homonuclear diatomic molecules. *J. Chem. Educ.*, **68**, 737 (1991).

69. W.J. JORGENSEN and L. SALEM, *The Organic Chemist Book of Orbitals*, Academic Press, USA (1973).

70. B.J. DUKE and B. O'LEARY, Non-Koopmans' molecules. *J. Chem. Educ.*, **72**, 501 (1995).

71. R.F.W. BADER, *Atoms in Molecules: A Quantum Theory*, in International Series of Monographs on Chemistry, Clarendon Press, UK (1995).

72. R.M. DREIZLER and E.K.U. GROSS, *Density-Functional Theory: An Approach to Quantum Many-Body Problems*, Springer, Berlin (1990).

73. L. PAULING, *The Nature of the Chemical Bond*, Cornell University Press, USA (1960).

74. R.S. MULLIKEN, A new electronegativity scale; together with data on valence states and valence ionization potentials and electron affinities. *J. Chem. Phys.*, **2**, 782 (1934).

75. A.L. ALLRED and E.G. ROCHOW, A scale of electronegativity based on electrostatic force. *J. Inorg. Nucl. Chem.*, **5**, 264 (1958).

76. L.C. ALLEN, Electronegativity is the average one-electron energy of the valence-shell electrons in ground-state free atoms. *J. Amer. Chem. Soc.*, **111**, 9003 (1989).

77. R.G. PARR and W. YANG, *Density-Functional Theory of Atoms and Molecules*, Oxford University Press, UK (1989).

78. R.G. PEARSON, Electronegativity scales. *Acc. Chem. Res.*, **23**, 1 (1990).

79. R.S. MULLIKEN, C.A. RIEKE, D. ORLOFF and H. ORLOFF, Overlap integrals and chemical binding. *J. Chem. Phys.*, **17**, 1248 (1949).

80. D.W. SMITH, A simple molecular orbital treatment of the barrier to internal rotation in the ethane molecule. *J. Chem. Educ.*, **75**, 907 (1998).

81. E. BESALÚ and J. MARTÍ, Exploring the Rayleigh–Ritz variational principle. *J. Chem. Educ.*, **75**, 105 (1998).

82. R. HOFFMANN, An extended Hückel theory. I. Hydrocarbons. *J. Chem. Phys.*, **39**, 1397 (1963).

83. J.A. POPLE and D.L. BEVERIDGE, *Approximate Molecular Orbital Theory*, McGraw Hill, USA (1970).

84. I. LEVINE, *Quantum Chemistry*, fourth edition, Allyn and Bacon, Boston, USA (1991).

85. R.F. HOUT Jr, W.F. PIETRO and W.J. HEHRE, *A Pictorial Approach to Molecular Structure and Reactivity*, John Wiley, UK (1984).

86. G.P. SHUSTERMAN and A.J. SHUSTERMAN, Teaching Chemistry with electron density models. *J. Chem. Educ.*, **74**, 771 (1997).

87. M. SPRINGBORG, Some recent density-functional studies of molecular systems. In *Density-Functional Methods in Chemistry and Materials Science*, edited by M. Springborg, John Wiley, UK (1997).

88. W.E. PALKE and B. KIRTMAN, Valence shell electron pair interactions in H_2O and H_2S: a test of the valence shell electron pair repulsion theory. *J. Am. Chem. Soc.*, **100**, 5717 (1978).

89. L.A.E. BAPTISTA DE CARVALHO and J.J. TEIXEIRA DIAS, Conformational analysis of dimethylethylamine: an *ab initio* m.o. study and comparison with ethylamine and ethylmethylamine. *J. Molec. Struct. (Theochem)*, **282**, 211 (1993).

90. G.N. LEWIS, The atom and the molecule. *J. Am. Chem. Soc.*, **38**, 762 (1916).

91. R.F.W. BADER, *Atoms in Molecules: A Quantum Theory*, in International-Series of Monographs on Chemistry, Clarendon Press, UK (1995) and references therein.

92. R.F.W. BADER, T.T. NGUYEN-DANG and Y. TAL, A topological theory of molecular structure. *Rep. Prog. Phys.*, **44**, 893 (1981).

93. R.F.W. BADER, T.S. SLEE, D. CREMER and E. KRAKA, Description of conjugation and hyperconjugation in terms of electron distributions. *J. Am. Chem. Soc.*, **105**, 5061 (1983).

94. D. CREMER and E. KRAKA, A description of the chemical bond in terms of local properties of electron density and energy. *Croatica Chem. Acta*, **57**, 1259 (1984).

95. A.W. POTTS and W.C. PRICE, The photoelectron spectra of methane, silane, germane and stannane. *Proc. Roy. Soc. Lond.*, **A326**, 165 (1972).

96. J. SIMMONS, Why equivalent bonds appear as distinct peaks in photoelectron spectra. *J. Chem. Educ.*, **69**, 522 (1992).

97. D.J. KLEIN and N. TRINAJSTIC, *Valence Bond Theory and Chemical Structure*, Elsevier, Amsterdam (1990) and references therein.

98. D.W. SMITH, The valence bond interpretation of molecular geometry. *J. Chem. Educ.*, **57**, 106 (1980).

99. R.P. FEYNMAN, Forces in molecules. *Phys. Rev.*, **56**, 340 (1939) and references therein.

100. D.P. BOYD and K.J. LIPKOWITZ, Molecular mechanics: the method and its underlying philosophy. *J. Chem. Educ.*, **59**, 269 (1982).

101. P.J. COX, Molecular mechanics: illustrations of its application. *J. Chem. Educ.*, **59**, 275 (1982).

102. K.J. LIPKOWITZ, Abuses of molecular mechanics. *J. Chem. Educ.*, **72**, 1070 (1995).

103. R.J. GILLESPIE and I. HARGITTAI, *The VSEPR Model of Molecular Geometry*, Prentice Hall, London, UK (1991) and references therein.

104. N.V. SIDGWICK and H.M. POWELL, Stereochemical types and valency groups. *Proc. Roy. Soc. Lond.*, **176A**, 153 (1940).

105. R.J. GILLESPIE, VSEPR method revisited. *Chem. Soc. Rev.*, **21**, 59 (1991).

106. R.J. GILLESPIE, The chemical bond, electron pair domains and the VSEPR model. *Chem 13 News*, **239**, (1995).

107. J.W. LINNETT, *The Electronic Structure of Molecules*, Methuen, UK (1964).

108. M. LAING, Shaping up with EAN and VSEPR. *Educ. in Chem.*, 102 (1995).

109. R.F.W. BADER, R.J. GILLESPIE and P.J. MACDOUGALL, A physical basis for the VSEPR model of molecular geometry. *J. Am. Chem. Soc.*, **110**, 7329 (1988).

110. M. LAING, No rabbit ears on water. *J. Chem. Educ.*, **64**, 124 (1987).

111. J.N. MURRELL, S.F.A. KETTLE and J.M. TEDDER, *The Chemical Bond*, John Wiley, USA (1978) and *Valence Theory*, second edition, John Wiley, USA (1970).

112. R.M. PITZER, Optimized molecular orbital wavefunction for methane constructed from a minimum basis set. *J. Chem. Phys.*, **46**, 487 (1967).

113. J. LENNARD-JONES, The molecular orbital theory of chemical valency, I. The determination of molecular orbitals. *Proc. Roy. Soc. Lond.*, **A198**, 1 (1949), and II. Equivalent orbitals in molecules of known symmetry. *Proc. Roy. Soc. Lond.*, **A198**, 14 (1949).

114. G.A. GALLUP, The Lewis electron-pair model, spectroscopy and the role of the orbital picture describing the electronic structure of molecules. *J. Chem. Educ.*, **65**, 671 (1988).

115. R.B. MARTIN, Localized and spectroscopic orbitals: squirrel ears on water. *J. Chem. Educ.*, **65**, 668 (1988).

116. L. PAULING, The nature of the chemical bond – 1992. *J. Chem. Educ.*, **69**, 519 (1992).

117. V.M.S. GIL and C.F.G. GERALDES, Uses and abuses of hybrid orbitals. *Rev. Port. Quím.*, **26**, 33 (1974).

118. R.F.W. BADER, M.T. CARROLL, J.R. CHEESEMAN and C. CHANG, Properties of atoms in molecules: atomic volumes. *J. Am. Chem. Soc.*, **109**, 7968 (1987).

119. M.L. DENNISTON, The generation of 2D and 3D electron density maps using high performance computing technology. *J. Chem. Educ.*, **70**, A76 (1993).

120. E. HÜCKEL, Quantumtheoretische Beitrage zum Benzolproblem, I. Die Elektronkonfiguration des Benzols und verwandter Verbindungen. *Z. Phys.*, **70**, 204 (1931).

121. A.A. FROST and B. MUSULIN, A mnemonic device for molecular orbital energies. *J. Chem. Phys.*, **21**, 572 (1953).

122. V.G.S. BOX and H.W. YU, π-Electron delocalization in organic molecules with C–N bonds. *J. Chem. Educ.*, **74**, 1293 (1998).

123. J.J.C. MULDER, The π-Electron system of monocylic polyenes $C_{2n}H_{2n}$ with alternating single and double bonds. *J. Chem. Educ.*, **75**, 594 (1998).

124. C.A. COULSON and G.S. RUSHBROOKE, Note on the method of molecular orbitals. *Proc. Camb. Phil. Soc.*, **36**, 193 (1940).

125. P. LASZLO, *La parole des choses*, Hermann editeurs des Sciences et des Artes, Paris (1993).

126. K.P. SUDLOW and A.A. WOOLF, What is the geometry at trigonal nitrogen? *J. Chem. Educ.*, **75**, 108 (1998).

127. N.K. KILDAHL, Bond energy data summarized. *J. Chem. Educ.*, **72**, 423 (1995).

128. R.S. TREPTOW, Bond energies and enthalpies: an often neglected difference. *J. Chem. Educ.*, **72**, 497 (1995).

129. R. PERKINS and C. LASSIGNE, What's the shape of that molecule? Part 1: a VSEPR short-cut. *Chem 13 News*, **217**, 4 (1995).

130. D.K. STRAUB, Lewis structures of oxygen compounds of 3p–5p nonmetals. *J. Chem. Educ.*, **72**, 889 (1995).

131. L. SUIDAN, J.K. BADENHOOP, E.D. GLENDENING and F. WEINHOLD, Common textbook and teaching misrepresentations of Lewis structures. *J. Chem. Educ.*, **72**, 583 (1995).

132. A.E. REED and P.V.R. SCHLEYER, Chemical bonding in hypervalent molecules. The dominance of ionic bonding and negative hyperconjugation over d-orbital participation. *J. Am. Chem. Soc.*, **112**, 1434 (1990).

133. O.J. CURNOW, A simple qualitative molecular-orbital/valence-bond description of the bonding in main group 'hypervalent' molecules. *J. Chem. Educ.*, **75**, 910 (1998).

134. P.G. NELSON, Valency. *J. Chem. Educ.*, **74**, 465 (1997).

135. D.K. STRAUB, Lewis structures of boron compounds involving multiple bonding. *J. Chem. Educ.*, **72**, 494 (1995).

136. V.M.S. GIL, S.J. FORMOSINHO and A.C. CARDOSO, Bond orders and multiple bonding. *Educ. in Chem.*, 11 (1988).

137. H.W. KROTO, J.R. HEATH, S.C. O'BRIEN, R.F. CURL and R.E. SMALLEY, C_{60}: Buckminsterfullerene. *Nature*, **318**, 162 (1985).

138. A.D.J. HAYMET, C_{120} and C_{60}: Archimidean solids constructed from sp^2 hybridized carbon atoms. *Chem. Phys. Lett.*, **122**, 421 (1985).

139. P.W. FOWLER and J. WOOLRICH, π-Systems in three dimensions. *Chem. Phys. Lett.*, **127**, 78 (1986).

140. H.P. LUTHI and J. ALMLOF, *Ab initio* studies of the thermodynamic stability of the icosahedral C_{60} molecule 'Buckminsterfullerene'. *Chem. Phys. Lett.*, **135**, 357 (1987).

141. R.L. DISCH and J.M. SCHULMAN, On symmetrical clusters of carbon atoms: C_{60}. *Chem. Phys. Lett.*, **125**, 465 (1986).

142. K. RAGHAVACHARI and C.M. ROHLFING, Structures and vibrational frequencies of C_{60}, C_{70} and C_{84}. *J. Phys. Chem.*, **95**, 5768 (1991).

143. D.W. BALL, Electronic absorptions of C_{60}: a quantum-mechanical model. *J. Chem. Educ.*, **71**, 463 (1994).

144. F. RIOUX, Quantum mechanics, group theory and C_{60}. *J. Chem. Educ.*, **71**, 464 (1994).

145. F. DE PROFT, C. VAN ALSENOY and P. GEELINGS, *Ab initio* study of the endohedral complexes of C_{60}, Si_{60} and Ge_{60} with monoatomic ions: influence of electrostatic effects and hardness. *J. Phys. Chem.*, **100**, 7440 (1996).

146. M. MARANGOLO, J. MOSCOVICI, G. LOUPIAS, S. RABII, S.C. ERWIN, C. HÉROLD, J.F. MARÊCHÉ and P. LAGRANGE, Experimental and theoretical study of the electron momentum density of K_6C_{60} and comparison to pristine C_{60}. *Phys. Rev.*, **B58**, 7593 (1998).

147. T.A. AILBRIGHT, J.K. BURDETT and M.-H. WHANGBO, *Orbital Interactions in Chemistry*, John Wiley, UK (1985).

148. J.G. VERKADE, Pictorialized solid state m.o.s: a nonmathematical but honest approach. *J. Chem. Educ.*, **68**, 739 (1991).

149. H. FUJIMOTO and K. FUKUI, *Chemical Reactivity and Reaction Paths*, edited by G. Klopman, Wiley-Interscience, USA (1974) and references therein.

150. J.L. REED, The Lewis structure: an expanded perspective. *J. Chem. Educ.*, **71**, 98 (1994).

151. R.B. WOODWARD and R. HOFFMANN, *The Conservation of Orbital Symmetry*, Academic Press, USA (1970).

152. J.A. POPLE and D.P. SANTRY, Molecular orbital theory of nuclear spin coupling constants. *Molec. Phys.*, **8**, 1 (1964).

153. M. KARPLUS, Contact electron–spin coupling of nuclear magnetic moments. *J. Chem. Phys.*, **30**, 11 (1959).

154. V.M.S. GIL and J.N. MURRELL, The relative importance of delocalization terms in proton spin–spin coupling constants. *Theor. Chim. Acta*, **4**, 114 (1966).

155. F.B. VAN DUIJNEVELDT, V.M.S. GIL and J.N. MURRELL, The calculation of directly bonded CH and CC coupling constants using delocalized molecular orbital theory. *Theor. Chim. Acta*, **4**, 85 (1966).

156. V.M.S. GIL, The relation of directly bonded CH coupling constants to s-characters revisited. *Theor. Chim. Acta*, **76**, 291 (1989).

157. M.D. MOSHER and S. OJHA, Hybridization and structural properties. *J. Chem. Educ.*, **75**, 888 (1998).

158. V.M.S. GIL and J.J.C. TEIXEIRA DIAS, Molecular orbital calculations of substituent effects on directly bonded CH coupling constants. *Molec. Phys.*, **15**, 47 (1968).

159. V.M.S. GIL and W. VON PHILIPSBORN, Electron lone-pair effects on nuclear spin coupling constants. *Magn. Reson. Chem.*, **27**, 409 (1989).
160. J.A. POPLE and A.A. BOTHNER-BY, Nuclear spin coupling between geminal H atoms. *J. Chem. Phys.*, **42**, 1339 (1965).
161. F. RIOUX and R.L. DEKOCK, The crucial role of kinetic energy in interpreting ionization energies. *J. Chem. Educ.*, **75**, 537 (1998).
162. T. CLARK and R. KOCH, *The Chemist's Electronic Book of Orbitals*, Springer, Berlin (1999).
163. D.R. LIDE (Ed.), *CRC Handbook of Chemistry and Physics*, 78th edition, CRC Press (1997–1998).

Index

Index